Basic Transport Phenomena in Materials Engineering

Manabu Iguchi • Olusegun J. Ilegbusi

Basic Transport Phenomena in Materials Engineering

Springer

Manabu Iguchi
Professor Emeritus
Graduate School
of Engineering
Division of Materials Science
and Engineering
Hokkaido University
Sapporo, Japan

Olusegun J. Ilegbusi
Professor
Department of Mechanical, Materials
and Aerospace Engineering
University of Central Florida
Orlando, Florida, USA

ISBN 978-4-431-54019-9 ISBN 978-4-431-54020-5 (eBook)
DOI 10.1007/978-4-431-54020-5
Springer Tokyo Heidelberg New York Dordrecht London

Library of Congress Control Number: 2013944231

Printed on acid-free paper

Springer is part of Springer Science+Business Media (www.springer.com)

Preface

Transport phenomena, involving fluid flow, heat transfer, and mass transfer, play a key role in the improvement of many materials processing operations. Examples include gas-stirred ladles and reactors, blast furnaces, excess carbon dioxide removal for an environmentally friendly steelmaking process, the floatation process, fuel cells, micro reactors, and the like. Optimization of these processes as well as development of new technology requires a fundamental understanding of the transport phenomena in such systems. This book is a concise compilation of basic information and data targeted at the practicing materials engineer and scientist, rather than an exhaustive treatment of the subject. It takes a holistic approach augmented with diagnostic example problems and solutions.

Transport phenomena in materials processing systems are inherently complex and difficult to characterize. The system could be single or multiphase, turbulent or laminar, steady or unsteady, or coupled involving concurrent multiple phenomena, and may cut across dimensional scales. Each of these complexities may in turn have peculiarities that require unique consideration. For example, a multiphase system may comprise continuous and dispersed phases with interphase transport of momentum, energy, and mass superimposed on the intraphase phenomena. Another is a process involving simultaneous fluid flow, heat transfer, and chemical reaction. The nature of these interactions is not yet fully understood and may require careful experimentation to augment theory. This book provides guidance on the relevant theory and experimental techniques required to understand such processes.

In order to improve the efficiency of conventional processes or to develop novel ones, a fundamental knowledge of transport phenomena therefore is highly desirable for researchers and engineers engaged in materials processing operations. Understanding the detailed transport phenomena in the reactor of materials processes based on experimentation alone is difficult at present for the following reasons: (i) the opacity of the material such as molten metals; (ii) the high temperature of the molten material such as steel at 1,600 °C; (iii) the complexity of turbulent flows where they occur; (iv) coupled heat and mass transfer in many cases; and (v) chemical reactions in certain situations. Mathematical modeling therefore is often required to complement and, sometimes, to validate the experimental findings. Formulating such mathematical models relies crucially on understanding the underlying fundamental principles that are introduced in this book.

Several new features are included in the book beyond the typical topics of traditional transport phenomena. It takes a holistic approach by considering both single and multiphase phenomena. There is a discussion on flow and heat transfer in microscale systems, a relevant issue in the design of modern materials processes including fuel cells. Both fundamental theory and experimental techniques are presented in the discussion on multiphase flow, as well as representation of auxiliary relationships such as turbulence models and interfacial phenomena.

This textbook is divided into four parts. Part I contains three chapters which describe momentum transfer in single-phase systems. This part also includes specialty topics such as rheology, flow in porous media, the role of surface forces, and microscale and nanoscale fluid mechanics. Part II has four chapters which focus on heat transfer in single-phase systems. It provides a brief discussion of the three modes of heat transfer, namely, conduction, convection, and radiation. Also included are phase transformation, heat transfer correlations, and natural and forced convection. Part III has one chapter covering introduction to diffusion and mass transfer, including processes occurring in homogeneous and porous media, and mass transfer involving chemical reaction. Part IV has three chapters which are devoted to momentum and heat transfer in multiphase flows. A variety of multiphase systems are considered including gas–liquid, solid–gas, and solid–liquid processes. Experimental techniques and results are also presented. Each of the four parts of the book includes example problems with solutions to assist student learning. In addition, diagnostic problems are provided to assess student comprehension of the material presented.

The subject of transport phenomena is quite extensive and the treatment here is by no means exhaustive. The reader is therefore referred to specialized texts on each of the subtopics for further details. The authors hope, however, that both graduate and undergraduate students will find this a useful introductory text on transport phenomena in materials processing operations as well as in the design and development of novel systems. Although the book is aimed primarily at students in materials science and engineering, it could also serve as a useful reference text for chemical engineering and mechanical engineering introductory courses in transport phenomena.

Sapporo, Japan Manabu Iguchi
Orlando, USA Olusegun J. Ilegbusi

Contents

Part II Single-Phase Heat Transfer

Nomenclature

Part 1

A	Cross-sectional area
A_p	Projected area
a, b	Height and width of a rectangular channel, respectively
C_D	Drag coefficient
C_f	Skin friction coefficient
c	Speed of sound
D_h	Hydraulic diameter
D	Pipe diameter
d_{ni}, d_{no}	Inner and outer diameters of nozzle, respectively
D_p, d_p	Particle diameter
Eo	Eötvös number
F	Force
F_B	Buoyancy force
F_D	Hydrodynamic drag
Fr	Froude number
Fr_m	Modified Froude number
F_w	Weight
f	Shedding frequency of Karma's vortex streets
g	Acceleration due to gravity
H	Height
H_w	Bath depth
h_L	Head loss
K	Bulk modulus of elasticity
Kn	Knudsen number
L	Representative length
L_e	Entrance length
M	Mach number
m	Area ratio
P	Peripheral length

p, p_0, p_g	Absolute, atmospheric and gauge pressures
p_d, p_s	Dynamic and static pressures, respectively
Q	Flow rate
Q_g, Q_L, Q_w	Gas flow rate, liquid flow rate, and water flow rate, respectively
R	Pipe radius
Re	Reynolds number
Re_c	Critical Reynolds number
St	Strouhal number
T	Temperature
Tu	Turbulence intensity
t	Time
V	Approaching flow velocity
V_B	Volume of bubble
V	Velocity
V_∞, v_∞	Terminal velocity
v_c	Centerline velocity
v^*	Friction velocity
W	Weight
We	Weber number
x, r, z	Cartesian coordinates
δ	Boundary layer thickness
λ	Friction coefficient
μ	Dynamic viscosity
ν_f	Kinematic viscosity
ρ	Density
σ	Surface tension
ξ	Coefficient
ζ	Loss coefficient

Part 2

A	Area (m^2)
d	Diameter (mm, cm, m)
F_τ	Viscous force (N)
Gr	Grashof number (−)
g	Acceleration due to gravity (m/s^2, cm/s^2)
K	Over-all heat transfer coefficient (W/m^2 K)
Nu	Nusselt number (−)
Nu_m	Mean Nusselt number (−)
Nu_x	Local Nusselt number (−)
Pr	Prandtl number (−)
Q	Heat flow (J)
q	Heat flux (W/m^2)
Ra	Rayleigh number (−)

Re_{L}	Mean Reynolds number (–)
Re_{x}	Local Reynolds number (–)
r	Radial distance (mm, cm, m)
t	Time (s)
x, y	Cartesian coordinates (mm, cm, m)
α	Heat-transfer coefficient (W/m^2 K)
α_{m}	Mean heat-transfer coefficient (W/m^2 K)
β	Volume coefficient of expansion (1/K)
$\Delta\theta$	Temperature difference (°C, K)
δ	Thickness of flat plate (mm, cm, m)
κ_{f}	Thermal diffusivity (m^2/s)
λ	Thermal conductivity (W/m K)
μ	Dynamic viscosity (Pa s)
ν_{f}	Kinematic viscosity (m^2/s)
θ	Temperature (°C, K)
θ_{f}	Temperature of fluid (°C, K)
θ_{c}	Contact angle (°)
θ_{s}	Temperature of solid body (°C, K)
ρ	Density (kg/m^3)
σ	Surface tension (N/m)
τ	Shear stress (Pa)

Part 3

A_r	Pre-exponential factor
C_{A}	Concentration of the diffusing species
c	Concentration of the diffusing species D_{AB} = binary diffusion coefficient.
$D_{\mathrm{A-B}}$	Diffusion coefficient
D_{AK}	Knudsen diffusion coefficient
D_{ij}	Binary diffusivities
$D_{i,m}$	Mass diffusion coefficient for species i in the mixture
$D_{T,i}$	Thermal diffusion coefficient
E_r	Activation energy for the reaction (J/kmol)
J_A	Molar flux of species A relative to the molar average velocity
$k_{f,r}$	Forward rate constant for reaction r
$k_{b,r}$	Backward rate constant for reaction r
k_T	Soret coefficient
L	Characteristic length
M_{A}, M_{B}	Molecular weights of species A and B respectively
$\mathcal{M}_i$	Symbol denoting species i
$M_{w,i}$	Molecular weight of species i
N	Number of chemical species in the system
$N_{i,}$ N_j	Fluxes

R	Universal gas constant (J/kmol)
R_A	Rate of production of species A by homogeneous chemical reaction
R_i	Net rate of production of species i by chemical reaction
$\widehat{R}_{i,r}$	Arrhenius molar rate of creation/destruction of species i in reaction r
S_i	Rate of reaction n by addition from the dispersed phase
U	Characteristic velocity
$v'_{i,r}$	Stoichiometric coefficient for reactant i in reaction r
x_A	Mole fraction of specy A
β_r	Temperature exponent
ε	Porosity
τ	Tortuosity

Part 4

A	Cross-sectional area of pipe
A_p	Projected area
A_{sB}	Surface area of bubble
b_f	Half-value radius of bubble frequency distribution
b_u	Half-value radius of liquid velocity distribution
b_α	Half-value radius of gas holdup distribution
c	Speed of sound
D	Vessel diameter
$D_{10}, D_{20}, D_{30}, D_{21}, D_{31}, D_{32}$	Mean diameters
D_h	Hydraulic diameter
d_B	Bubble diameter
d_{ni}, d_{no}	Inner and outer diameter of nozzle
d_{Bs}	Surface area equivalent diameter
E_0	Eötvös number
Fr_m	Modified Froude number
$f_{B,cl}$	Centerline value of bubble frequency
f_{BF}	Frequency of bubble formation
g	Acceleration due to gravity
H	Micro-channel height
j_G	Superficial velocity of gas
j_g	Superficial velocity of gas
j_L	Superficial velocity of liquid
j_T	Total superficial velocity $= j_G + j_L$
L	Representative length
M	Mach number
Mo	Morton number
Nu_{mp}	Mean Nusselt number
Pr	Prandtl number

Q_G	Gas flow rate
Q_g	Gas flow rate
Q_L	Liquid flow rate
Re	Reynolds number
Re_p	Particle Reynolds number
R	Pipe radius
Sc	Schmidt number
Sh	Sherwood number
Sh_m	Mean Sherwood number
T	Measurement time
Tu	Turbulence intensity
t_{Bi}	Residence time of *i*th bubble
u	Liquid velocity
u_g	Mean bubble rising velocity
$u_{B,max}$	Maximum bubble velocity
u_{cl}	Centerline value of liquid velocity
u_r	Representative velocity
V	Representative velocity
V_B	Volume of bubble
V_{LS}	Superficial velocity of liquid
V_{GS}	Superficial velocity of gas
v_n	Velocity of gas in single-hole nozzle
We	Weber number
We_{GL}	Weber number based on superficial velocity of liquid
We_{GS}	Weber number based on superficial velocity of gas
z	Axial distance measured from the nozzle exit
z_0, z_{0c}	Axial position at which gas holdup is 50 %
z_5	Axial position at which gas holdup is 5 %
α	Gas holdup
α_{cl}	Centerline value of gas holdup
β	Contraction ratio
μ	Dynamic viscosity
θ_a	Advancing contact angle
θ_c	Equilibrium contact angle
θ_r	Receding contact angle
ρ	Density
$\Delta\rho$	Density difference $= \rho_L - \rho_g$
ρ_g, ρ_L	Densities of gas and liquid, respectively
σ	Surface tension
ν_f	Kinematic viscosity of fluid
ν_L	Kinematic viscosity of liquid

Part I
Momentum Transfer in Single-Phase Flows

Chapter 1
Basic Properties of Fluid Flow

Abstract The real fluids can be classified into Newtonian and non-Newtonian fluids. Detailed explanation is given on the basic properties of Newtonian fluids such as air, water, and molten metal. The properties include the density, compressibility, and viscosity.

Keywords Compressibility • Density • Fluid • Gas • Liquid • Newtonian fluid • Non-Newtonian fluid • Pressure • Viscosity

1.1 Introduction

Materials processing typically involves transport phenomena in the fluid state. Most prevalent of such fluids are molten metals, semisolid alloy slurries in composite processing, water solution, etc. Molten metals can be classified as Newtonian in which the shear stress is proportional to the rate of deformation (velocity gradient). They, however, typically exist at very high temperatures compared to air and water. Some materials like molten slag and slurries sometimes behave as non-Newtonian fluid for which the above proportional relationship does not hold. Fortunately, most fluids in materials engineering exhibit Newtonian behavior and, hence, theoretical and experimental analyses of their flow phenomena are relatively simple. However, the boundary conditions at the wall in materials processing operations require particular attention. For example, the refractory used for reactors are usually poorly wetted by molten metals to prevent contamination of the metals due to undesirable chemical reactions. The no-slip boundary condition in wetted walls does not hold on a poorly wetted wall on which a liquid slips. Quantitative evaluation of the slip velocity has not yet been achieved.

In the beginning of this Part I of the book, the basic characteristics of a Newtonian fluid are described. The governing equations (equation of continuity, Navier–Stokes equation, and Bernoulli theorem) are thereafter introduced. In addition, the relationships are presented between pressure drop and cross-sectional

M. Iguchi and O.J. Ilegbusi, *Basic Transport Phenomena in Materials Engineering*,
DOI 10.1007/978-4-431-54020-5_1,

mean velocity in pipes, ducts, and some pipe elements due to their practical relevance to the design of pipeline systems. A method is also presented for estimating the hydrodynamic drag on a solid body immersed in a uniform approaching flow. Finally, the behaviors are discussed of non-Newtonian fluids, flow in porous media, and flow in microscale and nanoscale reactors. Additional information on these subjects should be referred to other texts [1–17].

1.2 Fluid Definition and Some Properties

1.2.1 Definition of a Fluid

Streeter and Wylie [10] proposed the following definition:

"A fluid is a substance that deforms continuously when subjected to a shear stress, no matter how small that shear stress may be."

Gases and liquids are classified into this category. For example, we can see in our daily life that water spreads readily in a vessel of any shape as soon as it is poured into that vessel. Solids on the other hand obey Hook's law. This law states that the deformation of a solid is proportional to the force acting on it in the elastic regime.

A fluid is regarded as a continuum under most conditions, implying that it is not necessary to consider the motions of individual molecules. This condition is quantitatively described by the Knudsen number, *Kn*, satisfying the following condition:

$$Kn = \frac{\lambda_m}{L} < \frac{1}{5} \quad \left(\lambda_m = \sum \lambda_i / N\right) \tag{1.1}$$

where λ_m (m) is the mean free path of molecules and L (m) is the representative length (see Fig. 1.1).

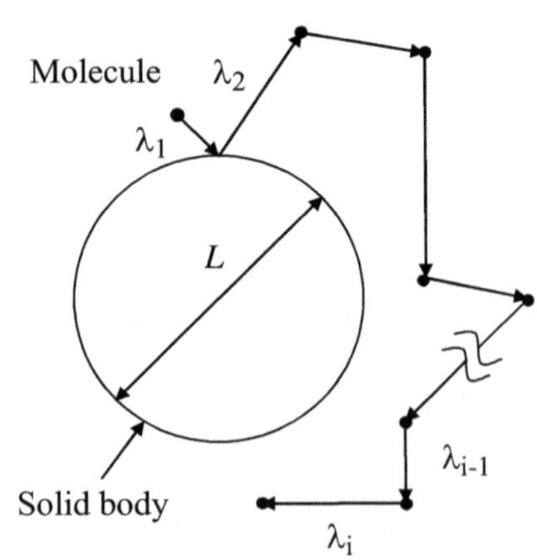

Fig. 1.1 Definition of mean free path

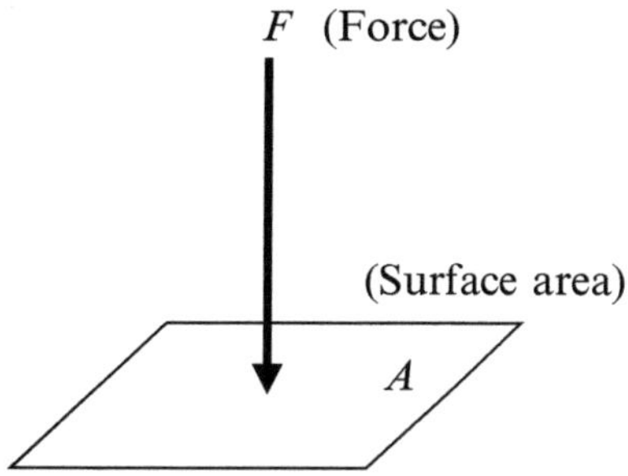

Fig. 1.2 Definition of pressure

1.2.2 Density

The density of a fluid ρ is defined as its mass per unit volume. The unit of density is (kg/m^3). The reciprocal of the density is called the specific volume, v_s (m^3/kg).

1.2.3 Pressure

The absolute pressure of a fluid p is defined as a force per unit area, as shown in Fig. 1.2:

$$p = F/A \tag{1.2}$$

where F is the force perpendicular to the surface and A is the surface area.

The unit of pressure is Pascal (Pa) or (N/m^2). It is often convenient in practical applications to measure the difference between the absolute pressure, p, and the atmospheric pressure, p_0. This pressure difference is called the gauge pressure, p_g. The relationship among the three pressures can be expressed by

$$p = p_0 + p_g \tag{1.3}$$

where $p_0 = 101.3$ kPa at 273 K.

Exercise 1.1 The depth of a lake, H, is 100 m, as shown in Fig. 1.3. Determine the absolute and gauge pressures, p and p_g, on the bottom of the lake. The density of water, ρ_L, is 997 kg/m^3 and the atmospheric pressure, p_0, is 101.3 kPa.

Answer to Exercise 1.1 As shown in Fig. 1.4, the balance of forces acting on a vertical water column of a cross-sectional area, A, is expressed by

$$pA = \rho_L gHA + p_0A \tag{1.4}$$

where g is the acceleration due to gravity. Dividing both sides of Eq. (1.4) by A gives

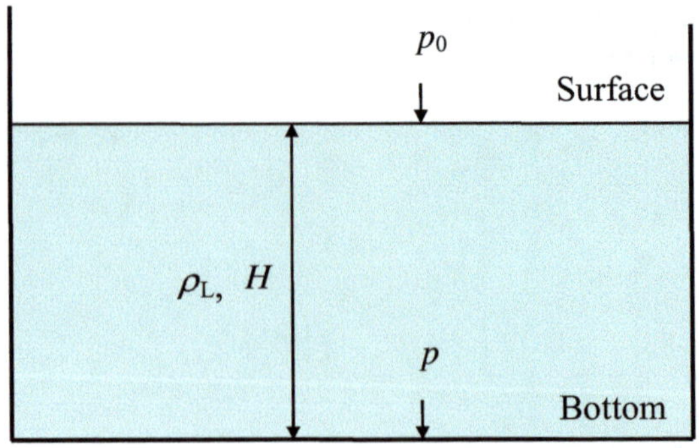

Fig. 1.3 Pressure on the bottom of a lake

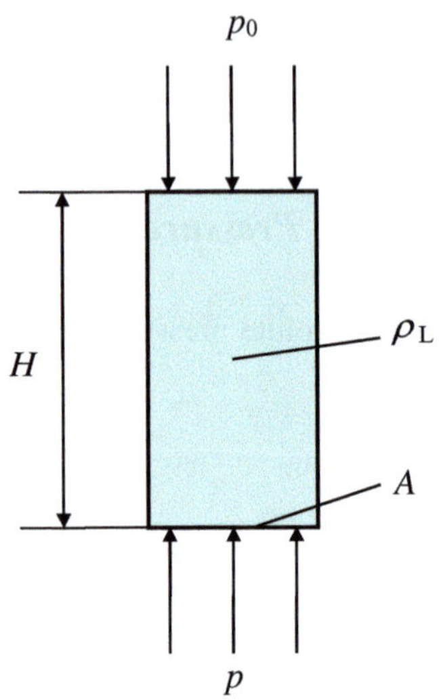

Fig. 1.4 Pressure acting on a vertical liquid column

$$p = \rho_L g H + p_0 \tag{1.5}$$

The gauge pressure, p_g, is

$$p_g = \rho_L g H \tag{1.6}$$

Substituting $\rho_L = 997\ \text{kg/m}^3$, $g = 9.80\ \text{m/s}^2$, and $H = 100$ m into Eq. (1.6) yields

$$p_g = 977.1\ \text{kPa} \tag{1.7}$$

Accordingly, the absolute pressure is given by

$$p = p_g + p_0 = 977.1\ \text{kPa} + 101.3\ \text{kPa} = 1{,}078.4\ \text{kPa} \tag{1.8}$$

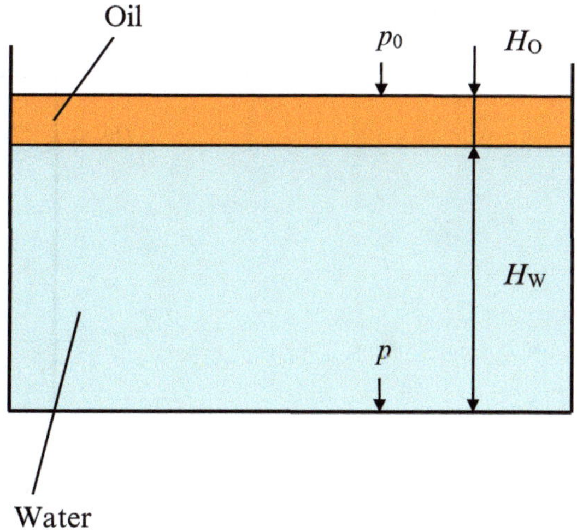

Fig. 1.5 Pressure on the bottom of a pond covered with oil

Exercise 1.2 The surface of a pond is covered with oil layer of thickness, $H_o = 2.0$ m. The original depth of the pond, H_w, is 10.0 m (see Fig. 1.5). Determine the absolute and gauge pressures, p and p_g, on the bottom of the pond. The densities of water and oil, ρ_w and ρ_o, are 997 kg/m^3 and 820 kg/m^3, respectively, and the atmospheric pressure, p_0, is 101.3 kPa.

Answer to Exercise 1.2 The gauge pressure, p_g, is expressed by

$$p_g = \rho_w g H_w + \rho_o g H_o \tag{1.9}$$

Substituting $\rho_w = 997$ kg/m^3, $\rho_o = 820$ kg/m^3, $g = 9.8$ m/s^2, $H_w = 10$ m, and $H_o = 2$m into Eq. (1.9) yields

$$\begin{aligned} p_g &= 997 \times 9.8 \times 10.0 + 820 \times 9.8 \times 2.0 \\ &= 97.7\,\text{kPa} + 16.1\,\text{kPa} = 113.8\,\text{kPa} \end{aligned} \tag{1.10}$$

The absolute pressure is given by

$$p = p_g + p_0 = 113.8 \text{ kPa} + 101.3 \text{ kPa} = 215.1 \text{ kPa} \tag{1.11}$$

Exercise 1.3 Figure 1.6 shows a U-shaped manometer used for measuring the pressure of water. The manometer is a device that employs liquid columns for determining differences in pressure [10]. The diameter of the manometer is uniform. Show that the pressure difference, $p_1 - p_2$, is expressed by

$$p_1 - p_2 = (\rho_{Hg} - \rho_w) g H \tag{1.12}$$

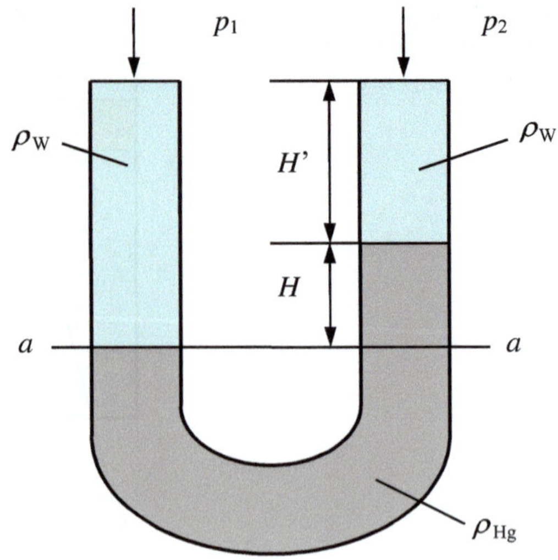

Fig. 1.6 U-shaped manometer

where ρ_{Hg} and ρ_w are the densities of mercury and water, respectively, g is the acceleration due to gravity, and H is the elevated length of the mercury column.

Answer to Exercise 1.3 A force balance on the horizontal plane denoted by *a–a* is given by

$$p_1 A + \rho_w g(H + H')A = p_2 A + \rho_w g H' A + \rho_{Hg} g H A \tag{1.13}$$

After some rearrangements Eq. (1.13) results in Eq. (1.11).

Exercise 1.4 The pressure in a liquid droplet is expressed by

$$p_{in} - p_{out} = 2\sigma/R \tag{1.14}$$

where σ (N/m) is the surface tension and R (m) is the radius of the droplet, as shown in Fig. 1.7. Derive Eq. (1.14). The surface tension is defined as force per unit length of the circumference of the droplet.

Answer to Exercise 1.4 Three forces acting on the droplet balance in the following manner (see Fig. 1.8):

$$p_{in}\pi R^2 = p_{out}\pi R^2 + 2\pi R\sigma \tag{1.15}$$

Dividing the left-hand and right-hand sides of Eq. (1.15) by πR^2 gives Eq. (1.14).

Exercise 1.5 Verify that the force due to the outer pressure p_{out} (see Fig. 1.7) is given by $\pi R^2 p_{out}$.

Answer to Exercise 1.5 Consider the force, dF, acting on an area element dA in Fig. 1.9, where dF and dA are expressed respectively as follows:

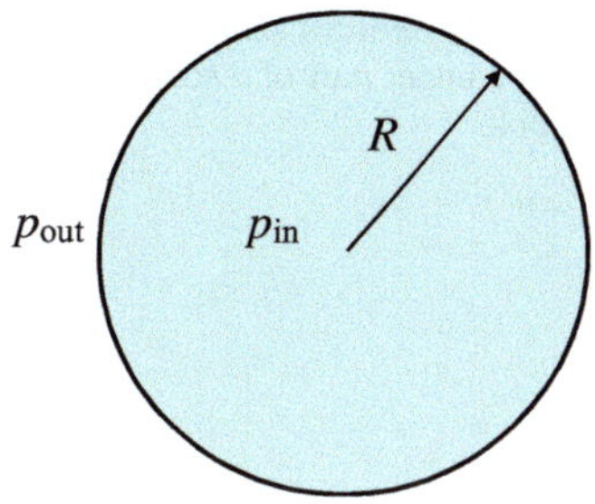

Fig. 1.7 Pressures outside and inside of a droplet

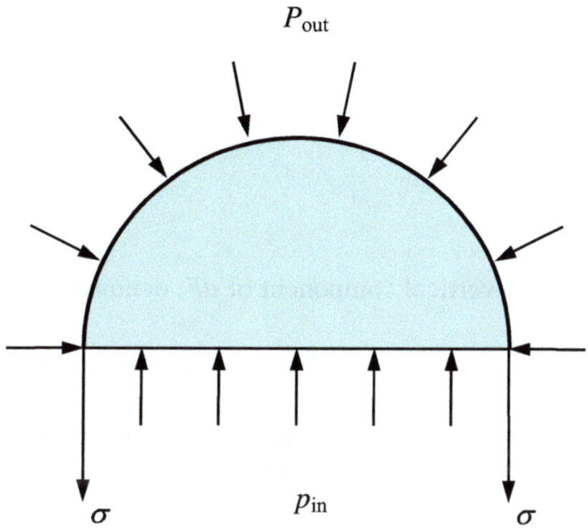

Fig. 1.8 Forces acting on the upper part of a droplet

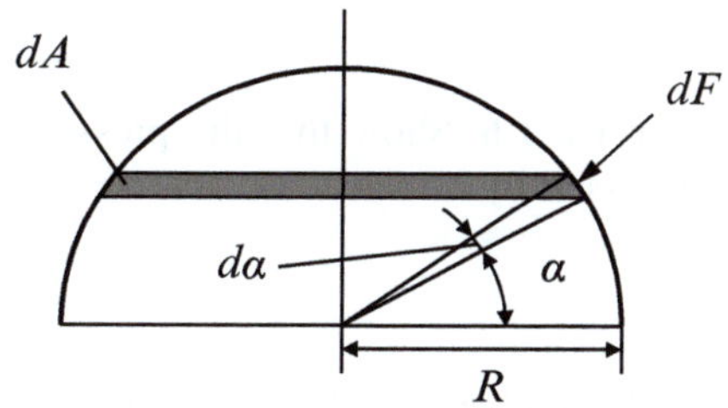

Fig. 1.9 Force acting on an area element on the surface of the upper part of a droplet

$$dF = p_{out} dA \tag{1.16}$$

$$dA = (R d\alpha) \times (2\pi R \cos\alpha) \tag{1.17}$$

Substituting Eq. (1.17) into Eq. (1.16) yields

$$dF = p_{out} \times (R d\alpha) \times (2\pi R \cos\alpha) = 2\pi R^2 p_{out} \cos\alpha d\alpha \tag{1.18}$$

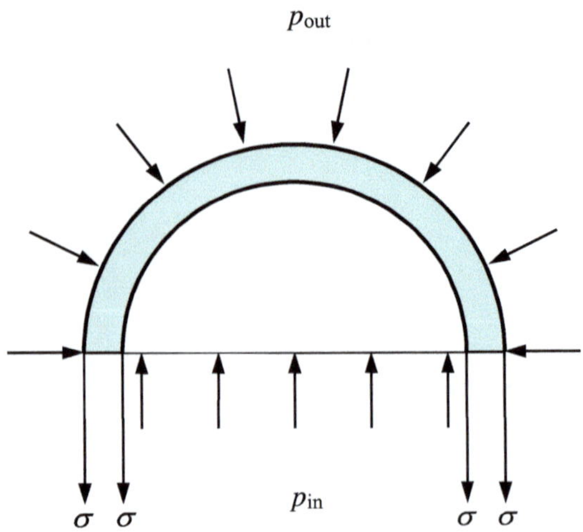

Fig. 1.10 Forces acting on the upper part of a soap bubble

The vertical component of dF, denoted by dF_v, is given by

$$dF_v = dF \sin\alpha = 2\pi R^2 p_{out} \cos\alpha \sin\alpha d\alpha \tag{1.19}$$

The total vertical force acting on the upper hemisphere can be obtained by integrating Eq. (1.19) from $\alpha = 0$ to $\pi/2$:

$$F_v = 2\pi R^2 p_{out} \int_0^{\pi/2} \cos\alpha \sin\alpha d\alpha = 2\pi R^2 p_{out} \times \frac{1}{2} = \pi R^2 p_{out} \tag{1.20}$$

Exercise 1.6 Show that the pressure in a soap bubble of inner diameter, $2R$, is expressed by

$$p_{in} - p_{out} = 4\sigma/R \tag{1.21}$$

The thickness of the bubble, t_{sb}, is assumed to be negligibly small in comparison with R.

Answer to Exercise 1.6 The balance of forces acting on the bubble (see Fig. 1.10) is given by

$$p_{in}\pi R^2 = p_{out}\pi R^2 + 2\pi R\sigma + 2\pi(R + t_{sb})\sigma \tag{1.22}$$

Dividing both sides of Eq. (1.22) by πR^2 and assuming that $t_{sb} << R$ yields Eq. (1.21). Accordingly, the pressure in a soap bubble is two times as large as that in a droplet.

1.2.4 Temperature

The common units of temperature are Kelvin (K), Celsius (°C), and Fahrenheit (°F). The units are related thus:

$$\theta(\mathrm{K}) = \theta(^\circ\mathrm{C}) + 273 \tag{1.23}$$

$$\theta(^\circ\mathrm{F}) = (9/5)\theta(^\circ\mathrm{C}) + 32 \tag{1.24}$$

1.3 Compressibility

1.3.1 Bulk Modulus of Elasticity, K

Gases are more compressible than liquids. The compressibility of a perfect gas is described by the perfect-gas law:

$$p = \rho RT \tag{1.25}$$

where p, ρ, and T are the absolute pressure, (Pa), density (kg/m^3), and temperature (K) of the gas, respectively, and R is the gas constant (N m/kg K).

The compressibility of a liquid is expressed by its bulk modulus of elasticity, K, thus:

$$K = -(dp/dV)V \tag{1.26}$$

where dp is the pressure increase, $-dV$ is the volume decrease, and V is the initial volume (see Fig. 1.11). The unit of K, (Pa), is the same as that of pressure, p.

Exercise 1.7 The pressure acting on water is increased from 100 kPa to 200 MPa. Calculate the volume decrease in (%) assuming that the bulk modulus of elasticity, K, is 2.0×10^3 MPa.

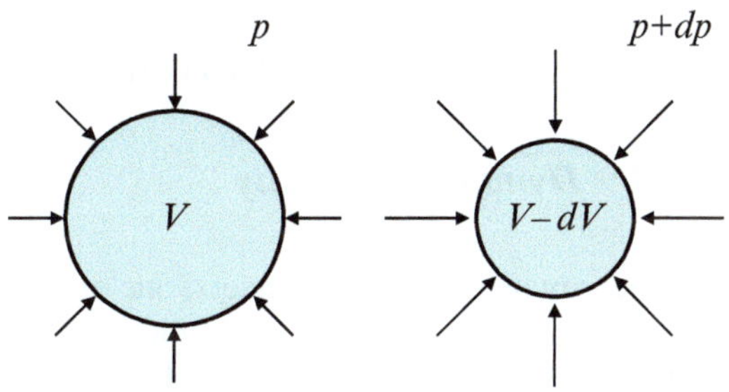

Fig. 1.11 Compressed droplet

Answer to Exercise 1.7 The volume decrease in (%) can be calculated from Eq. (1.26) as follows:

$$\begin{aligned} -100dV/V &= 100dp/K \\ &\approx 100\Delta p/K \\ &= 100 \times (200 \times 10^6 - 100 \times 10^3)/(2.0 \times 10^3 \times 10^6) = 10.0(\%) \end{aligned} \tag{1.27}$$

1.3.2 Speed of Sound

The speed of sound, c (m/s), in a fluid is closely associated with the compressibility of the fluid. Water is more difficult to compress than air, and the speed of sound in water (approximately 1,500 m/s) is much higher than that in air (approximately 340 m/s).

1.3.3 Mach Number

The Mach number, M (–), is defined as

$$M = v/c \tag{1.28}$$

where v (m/s) is the velocity of fluid and c (m/s) is the speed of sound at the fluid temperature. In a steady flow gas can be considered incompressible for $M < 0.3$. A steady liquid flow is considered to be incompressible under usual conditions. It should, however, be noted that the compressibility of a fluid plays an important role in unsteady flow of both gas and liquid. For example, water hammer and oil hammer occur due to sudden valve closure in pipeline systems. For simplicity, flow analysis often assumes the fluid is incompressible. The speed of sound in an incompressible fluid is infinite (∞).

1.4 Newton's Law of Viscosity

1.4.1 Dynamic Viscosity

Consider two concentric cylinders, one of which is slightly smaller than the other. The inner cylinder is fixed while the outer one is allowed to move, as shown in Fig. 1.12. A liquid fills the spacing between the cylinders and the outer cylinder is

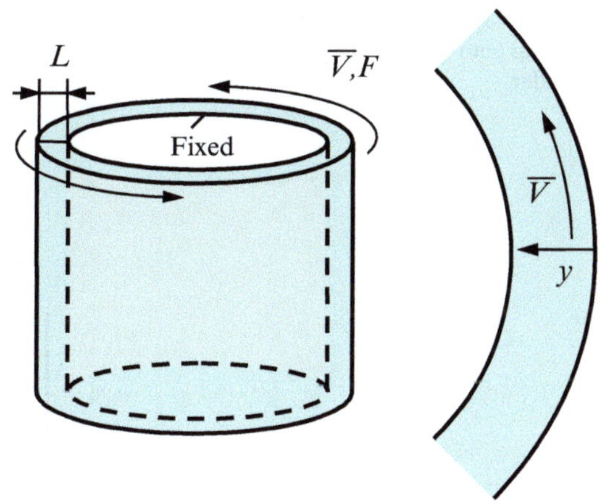

Fig. 1.12 Motion of a liquid contained in a gap between two concentric cylinders

rotated around its axis at a constant rotation velocity of V. The force required to rotate the vessel, F, is expressed by

$$F \propto VA/L \tag{1.29}$$

where A is the surface area of the inner wall of the outer cylinder wetted by the liquid and L is the clearance between the two cylinders.

Equation (1.29) is rewritten as

$$F/A = \mu V/L \tag{1.30}$$

where F/A is the shear stress, τ (N/m^2), and μ, the constant proportionality, is the dynamic viscosity. Equation (1.30) can be further rewritten in the following differential form:

$$\tau = -\mu dV/dy \tag{1.31}$$

where y is measured from the outer wall towards the centerline of the cylinders. Equation (1.31) is Newton's law of viscosity that relates the shear stress, τ, to the rate of deformation (velocity gradient) in the fluid. The minus sign is inserted on the right-hand side of Eq. (1.31) to ensure that the shear stress is positive with increasing wall distance y in Fig. 1.12 since the velocity gradient, dV/dy, is always negative in this situation.

1.4.2 Newtonian and Non-Newtonian Fluid

A fluid that obeys Newton's law of viscosity of Eq. (1.31) is called a Newtonian fluid. Air, water, oil, and molten metal are classified into this category. On the other hand,

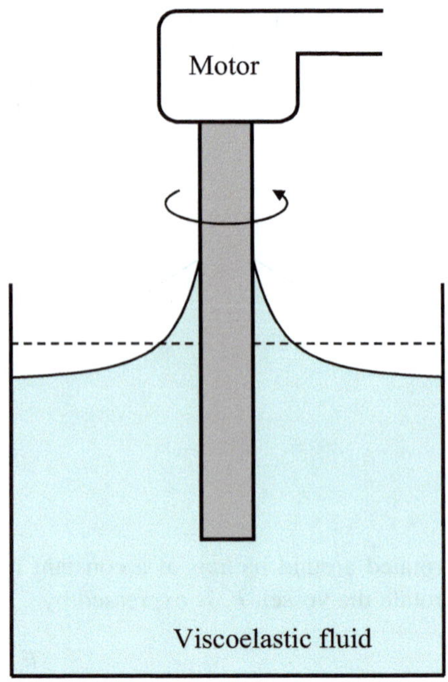

Fig. 1.13 Viscoelastic fluid clinging onto a rotating cylinder

a fluid which does not obey Eq. (1.31) is called a non-Newtonian fluid. Examples of non-Newtonian fluids include blood, semisolid slurries, and colloidal solutions.

A viscoelastic fluid is a non-Newtonian fluid that has both viscous and elastic nature and exhibits a very interesting characteristic. When a cylindrical rod is immersed vertically in a cylindrical bath and then rotated around the bath axis, a viscoelastic fluid clings to the rod and rises along it, as shown in Fig. 1.13. This phenomenon is known as the Weissenberg effect. Viscoelastic fluids are also considered to be shear thinning, implying that the viscosity decreases with the shear rate. A category of non-Newtonian fluids are stress thickening in which viscosity increases with increase of shear rate (see subsequent Fig. 3.1 later). An example of this category is a water–sand mixture.

Exercise 1.8 Why is the outer cylinder rotated and the inner cylinder kept stationary in the description of Newton's law of viscosity?

Answer to Exercise 1.8 Consider a Newtonian fluid between two concentric cylinders. When the outer vessel is rotated at a constant tangential velocity, $\overline{V}$, the velocity decreases linearly from v to 0 in the y direction. The force, F, is expressed by Eq. (1.30) under this condition. On the other hand, when the outer vessel is fixed and the inner vessel is rotated, the so-called Taylor vortex appears under specific conditions due to the centrifugal force acting on the liquid and the Taylor vortex pattern depends on the tangential velocity, $\overline{V}$, in a complex manner. Consequently, the velocity distribution does not always exhibit a linear distribution with respect to y. This situation is not suitable for measuring the dynamic viscosity, μ.

1.5 Surface Energy

1.5.1 Surface Tension and Contact Angle

A liquid has a natural tendency to minimize its surface area due to the surface tension, σ. Water has a surface tension of 73 (mN/m). The corresponding values for mercury and molten steel are approximately 460 mN/m and 1,700 (mN/m), respectively.

The wettability of a liquid droplet placed on a solid plate is quantitatively described by the contact angle, θ_c, as shown in Figs. 1.3, 1.4, and 1.5. In a strict sense, although θ_c is called the equilibrium contact angle, "equilibrium" is conventionally omitted throughout this book for brevity. A solid is wetted by a liquid for $0° \leq \theta_c < 90°$ and poorly wetted for $90° \leq \theta_c \leq 180°$. The refractory used in the steelmaking processes is typically poorly wetted by molten steel. The contact angle, θ_c, ranges approximately from 130° to 150° in order to avoid contamination of the molten steel from undesirable metallurgical reactions with the refractory.

In the case of a gas, when the Knudsen number (*Kn*) is less than 1/5, the gas velocity on the surface of a solid body is regarded as being equal to the velocity of the body. That is, there is no slip between the gas velocity on the surface of the body and the solid body velocity. This condition is known as the no-slip condition. On the other hand, the no-slip condition for a liquid is valid when a solid body is wetted by the liquid. When a solid body is poorly wetted by a liquid, some investigators have reported the existence of a slip velocity, v_s, on the solid body surface. However, it is not clear at present whether or not the slip velocity always exists under poor wettability conditions. The relationship between the contact angle, θ_c, and the slip velocity, v_s, is also unknown.

In this chapter we consider only cases for which the no-slip condition holds on the solid surface.

1.5.2 Capillary Force

Consider a system in which a small circular pipe is immersed vertically in a liquid bath. The liquid ascends in the pipe provided that the pipe wall is wetted by the liquid, as schematically illustrated in Fig. 1.14a. The elevation of the liquid in the pipe is given by

$$H = 4\sigma \cos \theta_c / (\rho g D) \tag{1.32}$$

This equation implies that a liquid descends in a poorly wetted pipe, as shown in Fig. 1.14b. The capillary force plays an important role in micro channels.

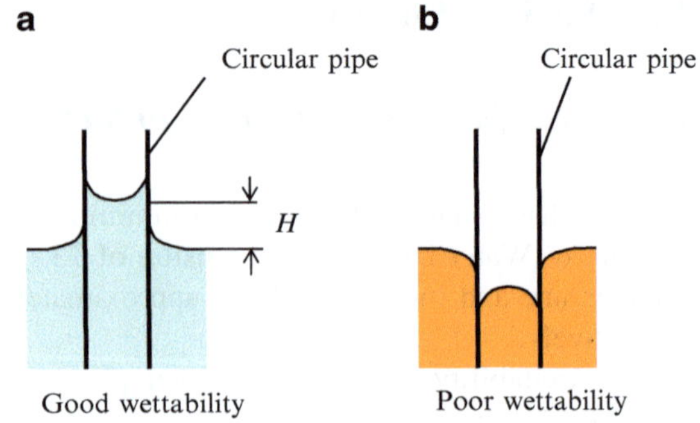

Fig. 1.14 Meniscus behaviors around wetted and poorly wetted cylinders. (**a**) Good wettability, (**b**) Poor wettability

Exercise 1.9 A glass pipe of inner diameter, D, of 2.0 mm is immersed vertically in a water bath. Determine the elevated height of water, H, in the pipe. The surface tension, σ, and the density, ρ, of water are 73 mN/m and 998 kg/m^3, respectively. The contact angle, θ_c, is 70°.

Answer to Exercise 1.9 Substituting $\sigma = 73$ mN/m, $\rho = 998$ kg/m^3, and $\theta_c = 70°$ into Eq. (1.32) gives

$$H = 5.1 \times 10^{-3}\ \text{m} = 5.1\ \text{mm} \tag{1.33}$$

References

1. Imai I (1979) Hydrodynamics. Shokabo Book Co., Ltd., Tokyo
2. Japan Society of Mechanical Engineers (1986) Handbook of mechanical engineering, A1 and A2. JSME, Tokyo
3. Japan Hydraulic and Pneumatics Society (1989) Handbook of hydraulics and pneumatic, New ed. Ohmusya Book Co., Ltd., Tokyo
4. Japan Society of Mechanical Engineers (1986) Handbook of mechanical engineering, A5, fluids engineering. JSME, Tokyo
5. Japan Society of Fluid Dynamics (1987) Handbook of fluid dynamics. Maruzen, Tokyo
6. Itaya M (1978) Fluid mechanics. Asakura Book, Co., Ltd., Tokyo
7. Mizushina A, Ogino F (1988) Transport phenomena. Sangyotosyo Book Co., Ltd., Tokyo
8. Hino M (1992) Hydrodynamics. Asakura Book, Co., Ltd., Tokyo
9. Tatsumi T (1983) Hydrodynamics. Baifukan Book, Co., Ltd., Tokyo
10. Streeter VL, Wylie EB (1985) Fluid mechanics, 8th edn. McGraw-Hill, New York
11. Ikui T, Inoue M (1982) Mechanics of viscous flow. Rikougakusya Book, Co., Ltd., Tokyo
12. Schlichting H (1979) Boundary-layer theory, translated by J. Kestin, 2nd edn. McGraw-Hill, New York
13. Uematsu T (1985) Fluid mechanics. Sangyotosyo Book Co., Ltd., Tokyo
14. Japan Society of Mechanical Engineers (1979) Mechanical engineering SI manual. JSME, Tokyo
15. Batchelor GK (2000) An introduction to fluid dynamics. Cambridge University Press, Cambridge
16. Bird RB, Stewart WE, Lightfoot EN (1960) Transport phenomena. Wiley, New York
17. Poirier DR, Geiger GH (1994) Transport phenomena in materials processing. TMS, Warrendale

Chapter 2
Momentum Transfer

Abstract The beginning of this chapter presents the governing equations for the motion of fluid flows: equation of continuity, equation of motion, and energy equation. In the following sections, dynamic similitudes and information required for designing pipeline systems are introduced.

Keywords Bernoulli equation • Equation of continuity • Equation of motion • Friction coefficient • Laminar flow • Navier–Stoke's equation • Pressure drop • Reynolds number • Transition to turbulence • Turbulent flow

2.1 Conservation Laws

The motion of a fluid is governed by conservation laws such as the equations of continuity, motion, and energy. These equations are collectively called the governing equations. Each equation will be described briefly in this section. Derivation of the governing equations is beyond the scope of this text, and the details are available in specialized texts including Schlichting [1], Bird et al. [2], and Poirier and Geiger [3].

2.1.1 Equation of Continuity

Figure 2.1 shows a fluid element in the Cartesian coordinate system. The conservation of mass for the fluid element is described by the equation of continuity. This equation is expressed for a Newtonian fluid as

$$\frac{\partial \rho}{\partial t}+\frac{\partial(\rho u)}{\partial x}+\frac{\partial(\rho v)}{\partial y}+\frac{\partial(\rho w)}{\partial z}=0 \tag{2.1}$$

M. Iguchi and O.J. Ilegbusi, *Basic Transport Phenomena in Materials Engineering*,
DOI 10.1007/978-4-431-54020-5_2, © Springer Japan 2014

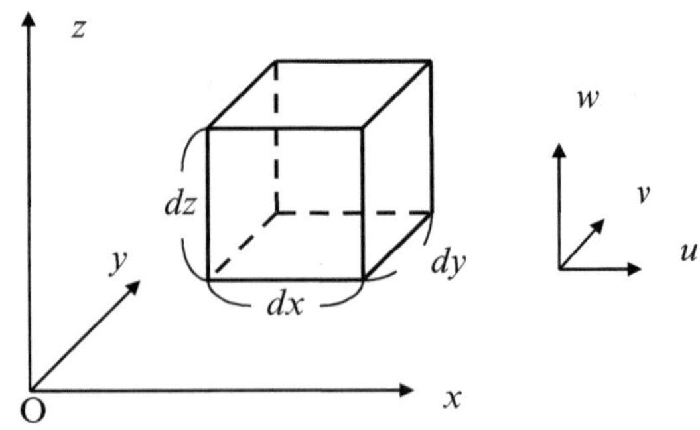

Fig. 2.1 Fluid particle in the Cartesian coordinate system

where ρ is the density of the fluid, t is time, and u, v, and w are the velocity components in the x, y, and z directions, respectively.

When the flow is steady, $\partial\rho/\partial t = 0$ and Eq. (2.1) becomes

$$\frac{\partial(\rho u)}{\partial x} + \frac{\partial(\rho v)}{\partial y} + \frac{\partial(\rho w)}{\partial z} = 0 \tag{2.2}$$

For a steady incompressible flow (ρ = constant), Eq. (2.2) reduces to

$$\frac{\partial u}{\partial x} + \frac{\partial v}{\partial y} + \frac{\partial w}{\partial z} = 0 \tag{2.3}$$

Exercise 2.1 Derive the equation of continuity, Eq. (2.1).

Answer to Exercise 2.1 Consider a fluid element at a position, P (x, y, z), as shown in Fig. 2.1. The mass of the fluid entering the element per unit time is expressed as

$$\rho u dydz + \rho v dzdx + \rho w dxdy \tag{2.4}$$

The outgoing mass of the fluid per unit time is given by

$$\begin{aligned}\left(u + \frac{\partial u}{\partial x}dx\right)\left(\rho + \frac{\partial \rho}{\partial x}dx\right)dydz + \left(v + \frac{\partial v}{\partial y}dy\right)\left(\rho + \frac{\partial \rho}{\partial y}dy\right)dzdx \\ + \left(w + \frac{\partial w}{\partial z}dz\right)\left(\rho + \frac{\partial \rho}{\partial z}dz\right)dxdy\end{aligned} \tag{2.5}$$

The mass accumulated in the element per unit time becomes

$$\frac{\partial}{\partial t}(\rho dxdydz) \tag{2.6}$$

The mass balance is expressed by

$$(\text{Accumulated mass}) = (\text{entering mass}) - (\text{outgoing mass}) \tag{2.7}$$

Substituting Eqs. (2.4)–(2.6) into Eq. (2.7) yields

$$\begin{aligned}\frac{\partial}{\partial t}(\rho dxdydz) = &\left[-\rho\frac{\partial u}{\partial x} - u\frac{\partial \rho}{\partial x} - \frac{\partial \rho}{\partial x}\frac{\partial u}{\partial x}dx\right]dxdydz \\ &+ \left[-\rho\frac{\partial v}{\partial y} - v\frac{\partial \rho}{\partial y} - \frac{\partial \rho}{\partial y}\frac{\partial v}{\partial y}dy\right]dxdydz \\ &+ \left[-\rho\frac{\partial w}{\partial z} - w\frac{\partial \rho}{\partial z} - \frac{\partial \rho}{\partial z}\frac{\partial w}{\partial z}dz\right]dxdydz\end{aligned} \tag{2.8}$$

The third terms within the parenthesis on the right-hand side of Eq. (2.8) are negligibly small compared to the other terms and, hence, Eq. (2.8) reduces after dividing both sides by *dxdydz* to

$$\frac{\partial \rho}{\partial t} = -\rho\frac{\partial u}{\partial x} - u\frac{\partial \rho}{\partial x} - \rho\frac{\partial v}{\partial y} - v\frac{\partial \rho}{\partial y} - \rho\frac{\partial w}{\partial z} - w\frac{\partial \rho}{\partial z} \tag{2.9}$$

The equation of continuity, Eq. (2.1), can readily be derived from this equation.

Batchelor [4] defined a streamline as a line in the fluid whose tangent is everywhere parallel to the velocity vector instantaneously. A stream tube is also defined as the surface formed instantaneously by all the streamlines at time t.

Streeter and Wylie [5] defined a streamline as a continuous line drawn through the fluid so that it has the direction of the velocity vector at every point. A stream tube is defined as the tube made by all the streamlines passing through a small, closed curve.

Irrespective of the definition, there is no through flow at the surface of a stream tube. Figure 2.2 shows a stream tube thus defined. The mass of a fluid entering the stream tube at a position 1 in time dt is expressed by

$$\rho v A dt \tag{2.10}$$

On the other hand, the mass of fluid flowing out of the tube at a position 2 is given by

$$\left(\rho + \frac{\partial \rho}{\partial s}ds\right)\left(v + \frac{\partial v}{\partial s}ds\right)\left(A + \frac{\partial A}{\partial s}ds\right)dt \tag{2.11}$$

The mass of fluid accumulated in the tube in time dt becomes

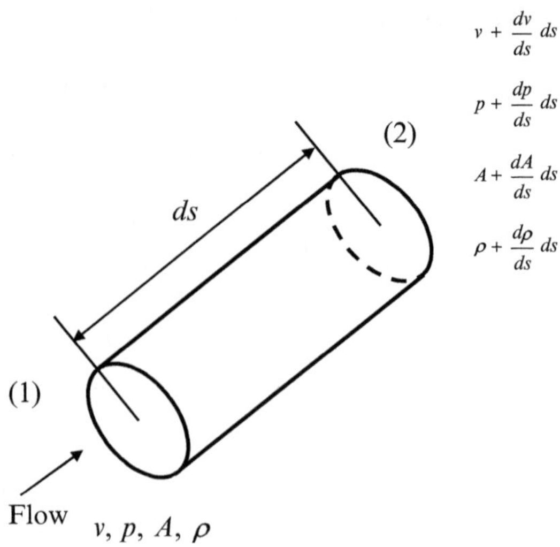

Fig. 2.2 Notations of velocity, v, pressure, p, cross-sectional area, A, and density, ρ, at the inlet and exit of a stream tube

$$\frac{\partial(\rho A ds)}{\partial t} dt \tag{2.12}$$

The mass balance in the stream tube is expressed by

$$\frac{\partial(\rho A ds)}{\partial t} dt = \rho v A dt - \left(\rho + \frac{\partial \rho}{\partial s} ds\right)\left(v + \frac{\partial v}{\partial s} ds\right)\left(A + \frac{\partial A}{\partial s} ds\right) dt \tag{2.13}$$

Rearranging Eq. (2.13) yields

$$\frac{\partial(\rho A)}{\partial t} + \frac{\partial(\rho v A)}{\partial s} = 0 \tag{2.14}$$

This equation is called the equation of continuity.

If a flow is steady ($\partial \rho A/\partial t = 0$), Eq. (2.14) becomes

$$\rho v A = \text{const.} \tag{2.15}$$

As the density of an incompressible fluid is constant, Eq. (2.14) reduces to

$$\frac{\partial A}{\partial t} + \frac{\partial(vA)}{\partial s} = 0 \tag{2.16}$$

If a flow is steady ($\partial A/\partial t = 0$), Eq. (2.16) is further simplified to

$$vA = \text{const.} \tag{2.17}$$

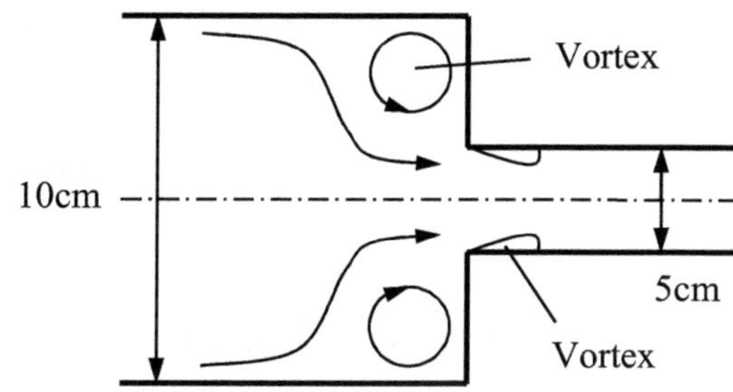

Fig. 2.3 Flow in an abrupt contraction

This relationship means that the flow rate Q ($=vA$) is kept constant in the flow direction in the stream tube when the flow is incompressible and steady.

In the case of real pipe flows, the no-slip condition holds on the pipe wall. It is therefore necessary to replace v in Eq. (2.17) by the cross-sectional mean velocity, v_m. The equation of continuity for compressible steady flow in a real pipe is therefore

$$\rho v_m A = \text{const.} \tag{2.18}$$

where ρ is the density of the fluid and A is the cross-sectional area of the pipe.

For an incompressible flow (ρ = constant), Eq. (2.18) reduces to

$$v_m A = Q = \text{const.} \tag{2.19}$$

The flow rate, Q, therefore is kept constant in the flow direction in real pipes.

Exercise 2.2 The diameter of a circular pipe, D, varies suddenly from 10.0 to 5.0 cm, as shown in Fig. 2.3. The cross-sectional mean velocity of a steady water flow in the larger pipe, v_{m1}, is 20 cm/s. Determine the cross-sectional mean velocity in the smaller pipe, v_{m2}.

Answer to Exercise 2.2 The equation of continuity, Eq. (2.18), is rewritten as

$$v_{m1} A_1 = v_{m2} A_2 \tag{2.20}$$

$$A_1 = \pi D_1{}^2/4 \tag{2.21}$$

$$A_2 = \pi D_2{}^2/4 \tag{2.22}$$

Combining Eqs. (2.20)–(2.22) gives

$$v_{m2} = v_{m1} (D_1/D_2)^2 \tag{2.23}$$

Substituting $v_{m1} = 20$ cm/s, $D_1 = 10.0$ cm, and $D_2 = 5.0$ cm into Eq. (2.23) yields

$$v_{m2} = 20 \times (10.0/5.0)^2 = 80 \text{ cm/s} \tag{2.24}$$

2.1.2 Navier–Stoke's Equation

The conservation of momentum for a compressible fluid flow was derived separately by Navier and Stokes. This equation therefore is usually called Navier–Stoke's equation. In the Cartesian coordinate system the components of Navier–Stoke's equation in the three directions for an incompressible fluid flow are expressed as follows:

x component

$$\rho\left(\frac{\partial u}{\partial t}+u\frac{\partial u}{\partial x}+v\frac{\partial u}{\partial y}+w\frac{\partial u}{\partial z}\right)=-\frac{\partial p}{\partial x}+\mu\left(\frac{\partial^2 u}{\partial x^2}+\frac{\partial^2 u}{\partial y^2}+\frac{\partial^2 u}{\partial z^2}\right)+F_x \tag{2.25}$$

y component

$$\rho\left(\frac{\partial v}{\partial t}+v\frac{\partial v}{\partial x}+v\frac{\partial v}{\partial y}+w\frac{\partial v}{\partial z}\right)=-\frac{\partial p}{\partial y}+\mu\left(\frac{\partial^2 v}{\partial x^2}+\frac{\partial^2 v}{\partial y^2}+\frac{\partial^2 v}{\partial z^2}\right)+F_y \tag{2.26}$$

z component

$$\rho\left(\frac{\partial w}{\partial t}+u\frac{\partial w}{\partial x}+v\frac{\partial w}{\partial y}+w\frac{\partial w}{\partial z}\right)=-\frac{\partial p}{\partial z}+\mu\left(\frac{\partial^2 w}{\partial x^2}+\frac{\partial^2 w}{\partial y^2}+\frac{\partial^2 w}{\partial z^2}\right)+F_z \tag{2.27}$$

where p is the static pressure, μ is the dynamic viscosity of fluid, and F_x, F_y, and F_z are the external forces acting on the fluid element in the x, y, and z directions, respectively.

Exercise 2.3 Show that the conservation of momentum reduces to the following equation for steady incompressible laminar flow between two parallel plates depicted in Fig. 2.4. The flow is assumed to be fully developed and steady and external forces are negligible.

$$-\frac{dp}{dx}+\mu\frac{d^2 u}{dy^2}=0 \tag{2.28}$$

Solution to Exercise 2.3 Since the flow is fully developed, the following relationships hold:

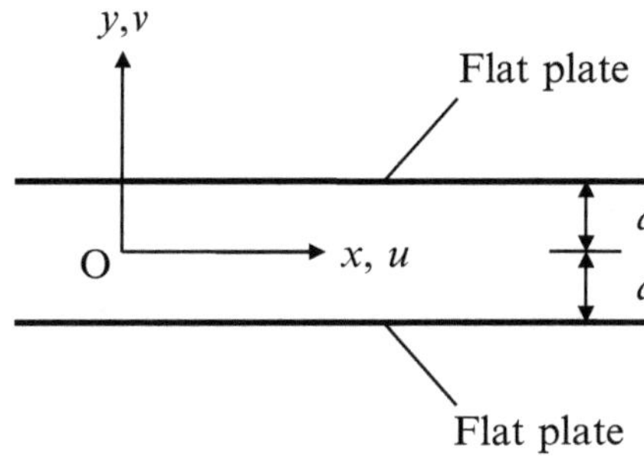

Fig. 2.4 Flow between two parallel flat plates

$$v = w = 0 \tag{2.29}$$

where v and w are the velocity components in the y and z directions, respectively.

$$\frac{\partial u}{\partial x} = \frac{\partial u}{\partial z} = 0 \tag{2.30}$$

Equation (2.30) implies that u is a sole function of y.

In addition, as the flow is steady and there is no external force,

$$\frac{\partial u}{\partial t} = \frac{\partial v}{\partial t} = \frac{\partial w}{\partial t} = 0 \tag{2.31}$$

$$F_x = F_y = F_z = 0 \tag{2.32}$$

Substituting these relationships into Eqs. (2.25)–(2.27) yields

x component

$$-\frac{\partial p}{\partial x} + \mu \frac{\partial^2 u}{\partial y^2} = 0 \tag{2.33}$$

y component

$$\frac{\partial p}{\partial y} = 0 \tag{2.34}$$

z component

$$\frac{\partial p}{\partial z} = 0 \tag{2.35}$$

Equations (2.34) and (2.35) collectively imply that the pressure, p, is a sole function of x.

Since u is a sole function of y and p is a sole function of x, Eq. (2.33) reduces to the following ordinary differential equation:

$$-\frac{dp}{dx}+\mu\frac{d^2u}{dy^2}=0 \tag{2.36}$$

In addition, the two terms in Eq. (2.36), $-dp/dx$ and $\mu d^2u/d^2y$, must be constant. The above equation may also be rewritten as

$$\mu\frac{d^2u}{dy^2}=\frac{dp}{dx} \tag{2.37}$$

Exercise 2.4 Solve Eq. (2.36) subject to the following boundary conditions:

$$u=0 \text{ at } y=\pm a \tag{2.38}$$

$$\frac{du}{dy}=0 \text{ at } y=0 \tag{2.39}$$

where $2a$ is the separation between the parallel passage.

Answer to Exercise 2.4 As mentioned above, the pressure gradient term, $-dp/dx$, is constant. Integrating Eq. (2.36) with respect to the vertical distance, y, yields

$$\mu\frac{du}{dy}=\frac{dp}{dx}y+C_1 \tag{2.40}$$

where C_1 is the integration constant. Equation (2.39) gives

$$C_1=0 \tag{2.41}$$

Further integration of Eq. (2.40) gives

$$\mu u=\frac{dp}{dx}\frac{y^2}{2}+C_2 \tag{2.42}$$

The integration constant, C_2, can be determined from Eq. (2.38) to be

$$C_2=-\frac{dp}{dx}\frac{a^2}{2} \tag{2.43}$$

Substituting Eq. (2.43) into Eq. (2.42) yields

$$u=-\frac{a^2}{2\mu}\frac{dp}{dx}\left[1-\left(\frac{y}{a}\right)^2\right] \tag{2.44}$$

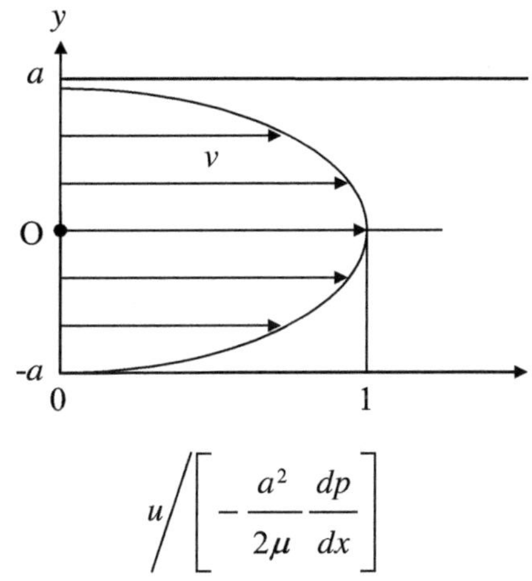

Fig. 2.5 Parabolic velocity distribution in a passage between two parallel flat plates

Equation (2.44) is rewritten as

$$\frac{u}{-\dfrac{a^2}{2\mu}\dfrac{dp}{dx}} = 1 - \left(\frac{y}{a}\right)^2 \tag{2.45}$$

Figure 2.5 shows that the velocity distribution is parabolic in shape.

2.2 Inviscid Flow and Bernoulli Equation

An inviscid fluid is a medium in which viscous effects can be considered negligible. An example is the outer flow region beyond the boundary layer region above a flat plate. The following equation holds for an incompressible, inviscid, steady flow in a stream tube (see Fig. 2.6). A stream tube is an imaginary tube considered in a fluid flow through whose wall the fluid does not cross.

$$\frac{p_1}{\rho g} + \frac{{v_1}^2}{2g} + z_1 = \frac{p_2}{\rho g} + \frac{{v_2}^2}{2g} + z_2 \tag{2.46}$$

where p_1 and v_1 are the pressure and velocity at a location 1. The vertical distance from that position to a reference position is denoted by z_1, and ρ and g are the density and the acceleration due to gravity, respectively. The corresponding values at location 2 are denoted by p_2, v_2, and z_2.

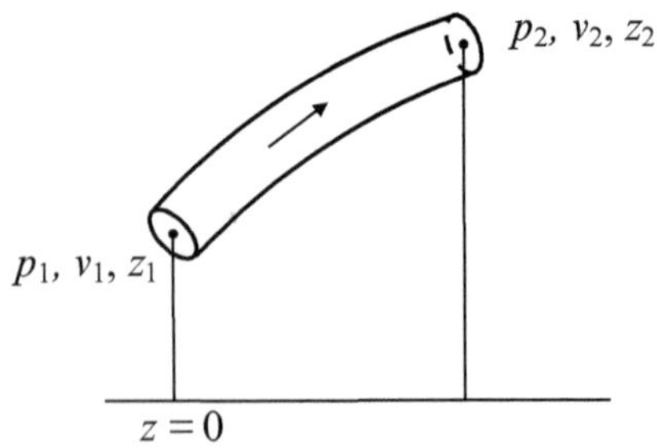

Fig. 2.6 Flow in a stream tube

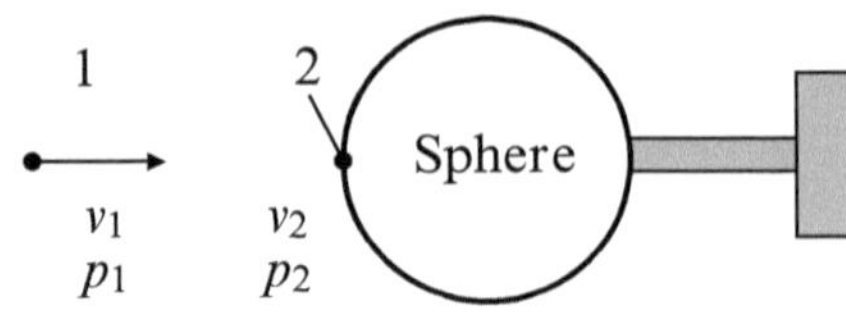

Fig. 2.7 Pressure on the forward stagnation point of a sphere fixed in uniform approaching flow

In the real pipeline systems Eq. (2.46) is modified as follows:

$$\frac{p_1}{\rho g} + \frac{{v_{m1}}^2}{2g} + z_1 = \frac{p_2}{\rho g} + \frac{{v_{m2}}^2}{2g} + z_2 + h_L \tag{2.47}$$

where v_{m1} and v_{m2} are the cross-sectional mean velocities at the locations 1 and 2, respectively, and h_L is the hydrodynamic loss. The method of evaluating h_L will be presented in a subsequent section.

Figure 2.7 shows a solid sphere immersed in a fluid flowing at velocity V. Location 1 is upstream of the sphere and location 2 is at the forward stagnation point of the sphere. Assuming $z_1 = z_2$, the Bernoulli equation becomes

$$\frac{p_1}{\rho g} + \frac{{v_1}^2}{2g} = \frac{p_2}{\rho g} + \frac{{v_2}^2}{2g} \tag{2.48}$$

Substituting $p_1 = p_s$, $v_1 = V$, $p_2 = p_t$, and $v_2 = 0$ into Eq. (2.48) gives

$$p_t = p_s + \frac{1}{2}\rho \bar{V}^2 \tag{2.49}$$

where p_t, p_s, and $\rho V^2/2$ are the total pressure, static pressure, and dynamic pressure, p_d, respectively.

Exercise 2.5 Water flows at a velocity of 3.0 m/s. Calculate the dynamic pressure, p_d, by assuming that the density of water, ρ, is 998 kg/m^3.

Answer to Exercise 2.5 The dynamic pressure, p_d, is given by

$$p_d = \rho V^2/2 = 998 \times 3.0 \times 3.0/2 = 4.49 \text{ kPa} \tag{2.50}$$

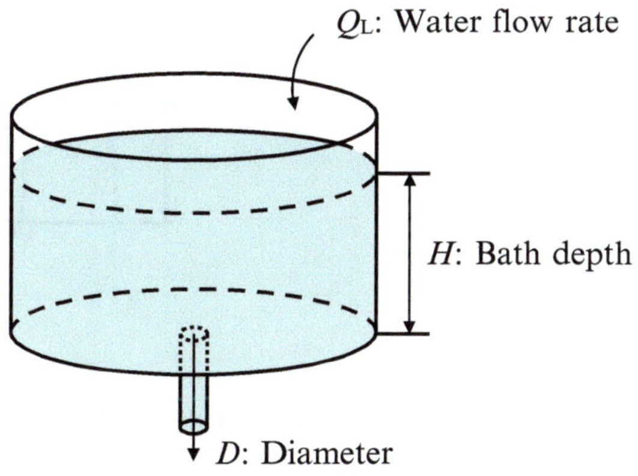

Fig. 2.8 Liquid discharge from a tank

Exercise 2.6 Water is poured into a very large cylindrical vessel at a flow rate, Q_L, of 2.0×10^{-2} m^3/s (see Fig. 2.8). The water issues through a hole at the bottom of the vessel. The diameter of the hole, D, is 5.0 cm. Calculate the depth of water, H, in the vessel at steady state by assuming that the hydrodynamic loss is negligible.

Answer to Exercise 2.6 Consider the flow between location 1 on the water surface and the location 2 at the exit of the bottom hole. The Bernoulli law without hydrodynamic loss is expressed by

$$\frac{p_1}{\rho g} + \frac{{v_{m1}}^2}{2g} + z_1 = \frac{p_2}{\rho g} + \frac{{v_{m2}}^2}{2g} + z_2 \tag{2.51}$$

Since the vessel is large, the falling velocity of the water surface, v_{m1}, is negligibly small, z_1 is equivalent to H, and $z_2 = 0$. The pressures, p_1 and p_2, are equal to the atmospheric pressure, p_0. Equation (2.51) reduces to

$$H = {v_{m2}}^2/(2g) \tag{2.52}$$

The cross-sectional mean velocity, v_{m2}, is

$$\begin{aligned} v_{m2} &= Q_L/(\pi D^2/4) = 2.0 \times 10^{-2} \times 4/\left[3.14 \times (5.0 \times 10^{-2})^2\right] \\ &= 10.2\ \mathrm{m/s} \end{aligned} \tag{2.53}$$

Substituting Eq. (2.53) into Eq. (2.52) yields

$$H = (10.2)^2/(2 \times 9.80) = 5.31\ \mathrm{m} \tag{2.54}$$

Exercise 2.7 Water flows in a horizontal circular pipe from location 1 to location 2, as shown in Fig. 2.9. The cross-sectional mean velocity at location 1 is 3.0 m/s, and the density of water is 1,000 kg/m^3. Calculate the difference between the gauge

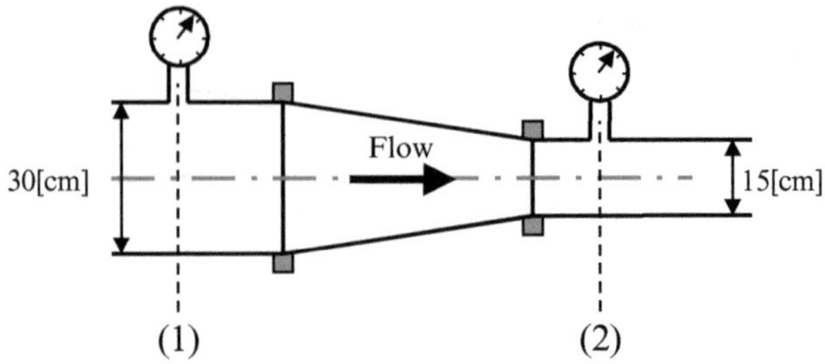

Fig. 2.9 Flow in a converging pipe

pressure at location 1, p_1, and that at location 2, p_2. The hydrodynamic loss is negligible.

Answer to Exercise 2.7 The cross-sectional mean velocity at location 2, v_{m2}, can be calculated from the equation of continuity, thus:

$$v_{m2} = v_{m1}(A_1/A_2) = v_{m1}(D_1/D_2)^2 = 3.0 \times (0.30/0.15)^2 = 3.0 \times 4 = 12\ \text{m/s} \tag{2.55}$$

The Bernoulli equation without hydrodynamic loss is

$$\frac{p_1}{\rho g} + \frac{{v_{m1}}^2}{2g} + z_1 = \frac{p_2}{\rho g} + \frac{{v_{m2}}^2}{2g} + z_2 \tag{2.56}$$

Since the pipe is placed horizontally,

$$z_1 = z_2 \tag{2.57}$$

Substituting Eq. (2.57) into Eq. (2.56) gives

$$p_1 - p_2 = \rho({v_{m2}}^2 - {v_{m1}}^2)/2 = 1{,}000 \times (144 - 9)/2 = 6.75 \times 10^4\ \text{Pa} \tag{2.58}$$

2.3 Flow Regime

2.3.1 Laminar–Turbulent Transition

More than 100 years ago, Reynolds systematically investigated flows in a circular pipe and observed that two types of patterns (laminar and turbulent flows) appear with respect to the pipe diameter, D, cross-sectional mean velocity, v_m, and the kinematic viscosity of fluid, ν_f $(=\mu/\rho)$. Finally, he derived the following dimensionless number to discriminate the two flow patterns:

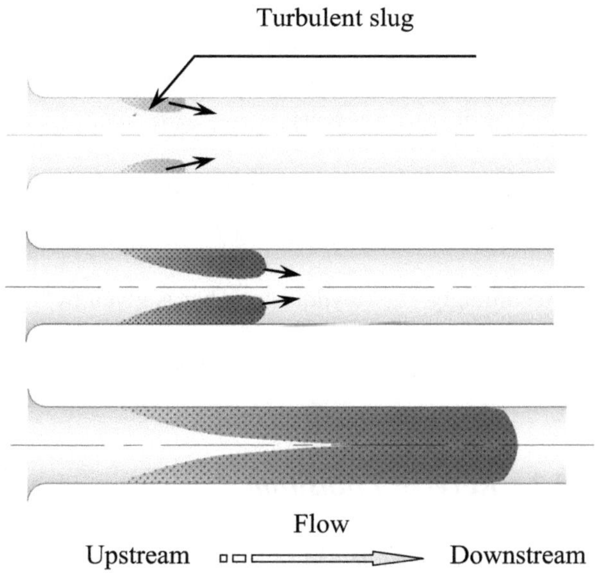

Fig. 2.10 Generation and propagation of a turbulent slug in a circular pipe

$$Re = v_m D / \nu_f \tag{2.59}$$

Later, this number was named the Reynolds number in honor of the excellent work of Reynolds. The kinematic viscosity is a measure of the momentum transfer rate in a fluid flow. The Reynolds number in effect is a measure of the ratio of inertial force to viscous force in the fluid flow.

Turbulent flow is characterized by many vortices of different sizes. These vortices are believed to be generated due to the bursting phenomenon. The bursting phenomenon is characterized by two typical motions called the sweep and ejection. The critical Reynolds number for transition from laminar flow to turbulent flow in a smooth circular pipe is

$$Re_c = 2{,}320 \tag{2.60}$$

Even if the disturbance in an incoming flow is very strong, the flow in the pipe is maintained laminar for $Re < 2{,}320$. This critical Reynolds number therefore is called the lower critical Reynolds number.

On the other hand, when the disturbance is very weak, laminar flow is maintained until the Reynolds number reaches another critical value greater than 2,320. This critical value is dependent on the magnitude of the disturbance and called the upper critical Reynolds number.

The lower critical Reynolds number in a noncircular pipe is different from that in a circular pipe and dependent on the shape of the cross section.

In the transition of Reynolds number range a turbulent slug appears almost periodically near the entrance of the pipe and then propagates in the downstream direction, as shown in Fig. 2.10.

Exercise 2.8 Water with density, ρ, of 997 kg/m^3 and kinematic viscosity, ν_f, of 1.00×10^{-6} m^2/s flows in a circular pipe 5.0 cm in diameter, D, at a flow rate, Q_w, of 2.0×10^{-2} m^3/min. Calculate the Reynolds number, *Re*.

Answer to Exercise 2.8 The cross-sectional mean velocity, v_m, is given by

$$v_m = 4Q_w/(\pi D^2) = 4 \times (2.0 \times 10^{-2}/60)/(3.14 \times 0.050 \times 0.050) \\ = 0.170 \text{ m/s} \tag{2.61}$$

Accordingly, the Reynolds number is given by

$$Re = v_m D/\nu_f = 0.170 \times 0.050/(1.00 \times 10^{-6}) = 8.5 \times 10^3 \tag{2.62}$$

2.3.2 *Reverse Transition*

A turbulent flow subjected to high spatial acceleration may become laminar. This phenomenon is called the reverse transition or relaminarization. This situation is observed, for example, in a nozzle with very high contraction ratio.

2.3.3 *Instantaneous Velocity*

Velocities typical of steady laminar and turbulent flows are schematically shown in Fig. 2.11. The velocity at a point in a laminar flow is independent of time, while that in a turbulent flow fluctuates around a mean value of $\bar{v}$. This is because of the many vortices that are present in the turbulent flow. As the flow is steady, $\bar{v}$ is constant. The deviation of the instantaneous velocity, v, from its mean value, $\bar{v}$, is designated the turbulent or fluctuating component, v'.

2.4 Flow in a Pipe

2.4.1 *Flow Development*

The flow velocity distribution in a pipe varies in the axial direction until the boundary layer developing on the pipe wall reaches the centerline of the pipe (see Fig. 2.12). The axial regions upstream and downstream of this position are called the developing region and the fully developed region, respectively. The developing region is sometimes called the entrance region or the inlet region.

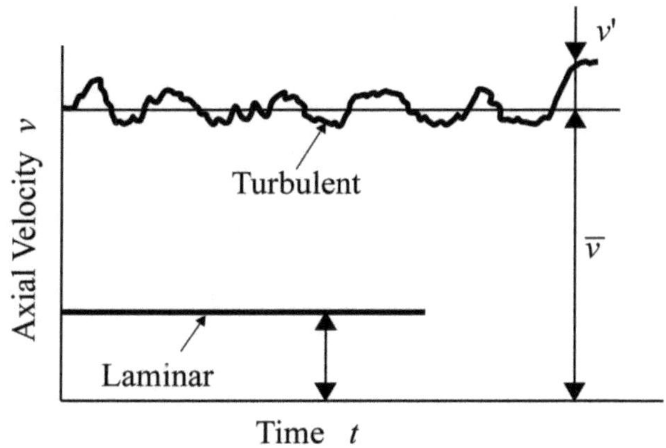

Fig. 2.11 Velocity signals in laminar and turbulent flows

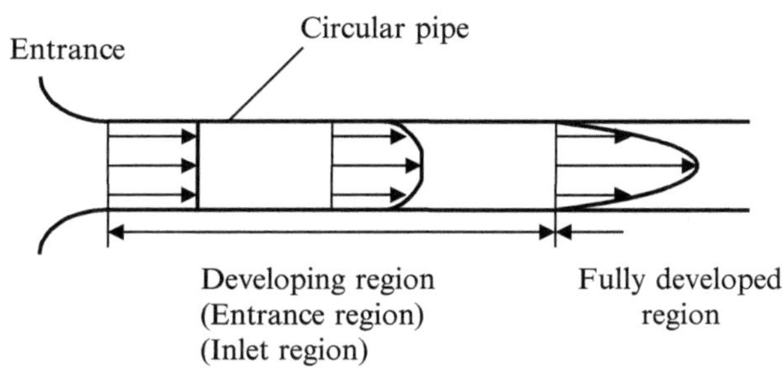

Fig. 2.12 Flow field near the entrance of a circular pipe

The entrance length, L_e, is introduced to quantitatively determine the extent of the developing region. It is defined as the distance from the entrance of the pipe to the axial position at which the centerline velocity, v_{cl}, becomes $0.99v_{cl,fd}$, where $v_{cl,fd}$ is the centerline velocity in the fully developed region. The entrance length of a flow in a circular pipe is given by

$$L_e = 0.05DRe \quad \text{(laminar)} \tag{2.63}$$

$$L_e = 50D \ \text{(turbulent)} \tag{2.64}$$

where Re is the Reynolds number expressed by Eq. (2.59). When the flow is turbulent, Eq. (2.64) is approximately valid for noncircular pipes such as a rectangular pipe. In the latter case D is interpreted as the hydraulic diameter defined subsequently in Sect. 2.4.5.

Exercise 2.9 Silicone oil of density 936 kg/m^3 and kinematic viscosity 100 mm^2/s is flowing in a horizontal circular pipe of inner diameter, D, of 1.0 cm. The flow rate, Q_L, is 1.5×10^{-3} m^3/min. Estimate the entrance length, L_e.

Answer to Exercise 2.9 The cross-sectional mean velocity, v_m, is given by

$$v_m = 4Q_L/(\pi D^2) = 4 \times (1.5 \times 10^{-3}/60) \Big/ \left[3.14 \times (1.0 \times 10^{-2})^2\right]$$
$$= 0.318 \text{ m/s} \quad (2.65)$$

The Reynolds number is calculated from Eq. (2.59) to give

$$Re = v_m D/\nu_f = 0.318 \times 1.0 \times 10^{-2}/(100 \times 10^{-6}) = 31.8 \quad (2.66)$$

This Reynolds number is much smaller than the critical Reynolds number of 2,320. Accordingly, the flow is laminar. Equation (2.63) therefore yields

$$L_e = 0.05DRe = 0.05 \times 1.0 \times 10^{-2} \times 31.8 = 0.016 \text{ m} \quad (2.67)$$

2.4.2 *Velocity Distribution*

2.4.2.1 Laminar Pipe Flow

The velocity distribution of laminar flow in a circular pipe is expressed by

$$\frac{v}{v_{cl}} = 1 - \left(\frac{r}{R}\right)^2 \text{(laminar)} \quad (2.68)$$

where v is the axial velocity component, v_{cl} is the centerline value of v, r is the radial distance, and R is the radius of the pipe (see Fig. 2.13).

The relationship between the cross-sectional mean velocity, v_m, and the centerline velocity, v_{cl}, is given by

$$v_m = v_{cl}/2 \quad (2.69)$$

2.4.2.2 Turbulent Pipe Flow

The following velocity distribution holds when flow in a circular pipe is turbulent, as shown in Fig. 2.14;

$$\frac{\overline{v}}{\overline{v}_{cl}} = \left(1 - \frac{r}{R}\right)^{1/7} \text{(turbulent)} \quad (2.70)$$

This velocity distribution is termed the 1/7th power law and it is valid for $3{,}000 < Re < 10^5$.

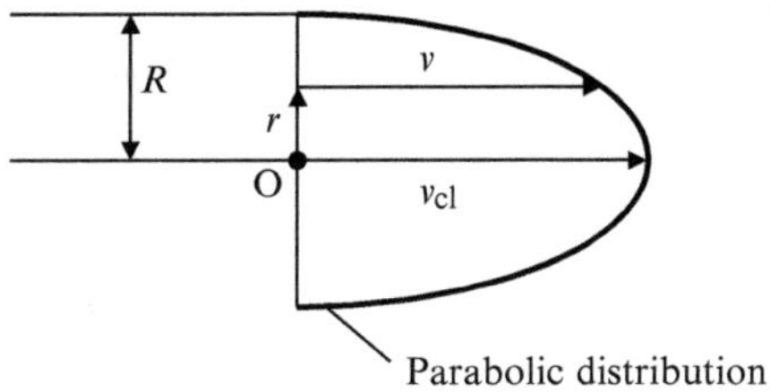

Fig. 2.13 Parabolic velocity distribution in a circular pipe

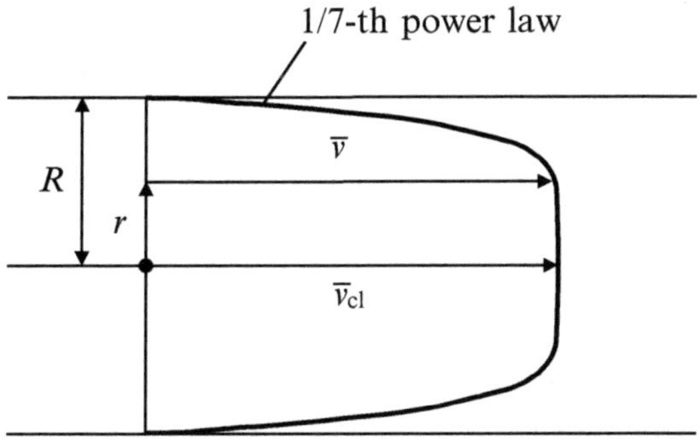

Fig. 2.14 1/7th-Power law velocity distribution in a circular pipe

The following 1/*n*th power law is valid over a wider Reynolds number range than the 1/7th power law:

$$\frac{\overline{v}}{\overline{v}_{\mathrm{cl}}} = \left(1 - \frac{r}{R}\right)^{1/n} \text{(turbulent)} \tag{2.71}$$

$$n = 2.0\log_{10}(Re/10)\ (3{,}000 < Re < 10^6) \tag{2.72}$$

These two velocity distributions can satisfactorily approximate the measured distributions in the central part of a circular pipe. However, the accuracy of approximation is insufficient near the pipe wall. Accordingly, the following logarithmic velocity distributions have been proposed (see Fig. 2.15):

$$\frac{\overline{v}}{v*} = y*\ (y* \leq 5) \tag{2.73a}$$

$$\frac{\overline{v}}{v*} = 11.5\log_{10} y* - 3.05\ (5 < y* < 70) \tag{2.73b}$$

$$\frac{\overline{v}}{v*} = 5.75\log_{10} y* + 5.5\ (70 \leq y*) \tag{2.73c}$$

$$v* = (\tau_{\mathrm{w}}/\rho)^{1/2} \tag{2.74}$$

$$y* = \frac{yv*}{v_{\mathrm{f}}} \tag{2.75}$$

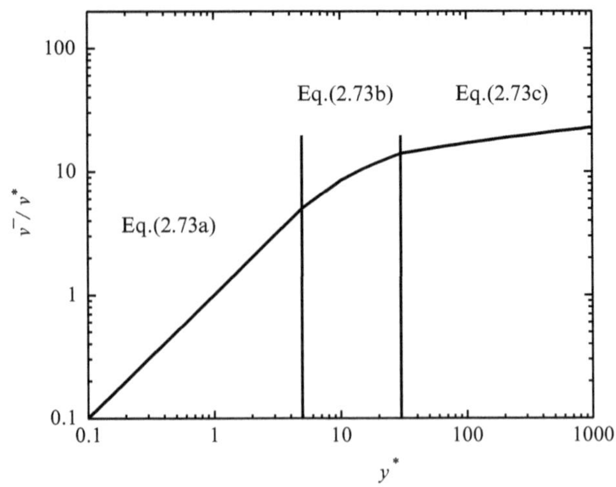

Fig. 2.15 Logarithmic velocity distribution in a circular pipe

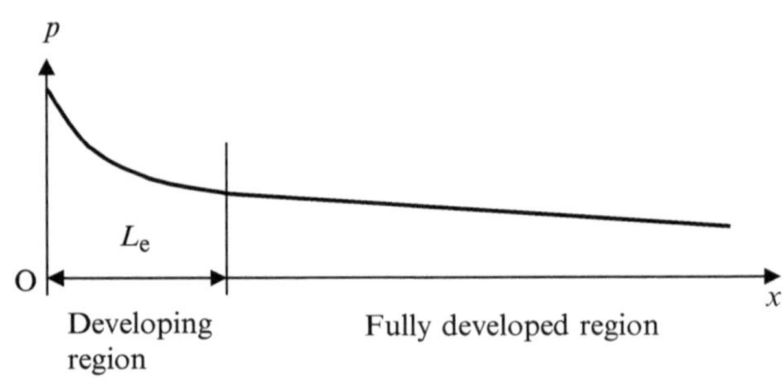

Fig. 2.16 Pressure distribution near the entrance of a circular pipe

where v^* is the friction (or shear) velocity, τ_w is the wall shear stress, and y (=$R - r$) is the distance from the wall.

2.4.3 Pressure Drop

The static pressure in a circular pipe first decreases rapidly in the developing region with axial distance from the entrance of the pipe and then decreases linearly with axial distance in the fully developed region (see Fig. 2.16). The pressure drop, Δp, in the fully developed region is expressed by

$$\Delta p = \lambda \frac{L}{D} \frac{\rho v_m^2}{2} \tag{2.76}$$

where λ is the friction coefficient, L is the pipe length, D is the pipe diameter, ρ is the density of the fluid, and v_m is the cross-sectional mean velocity.

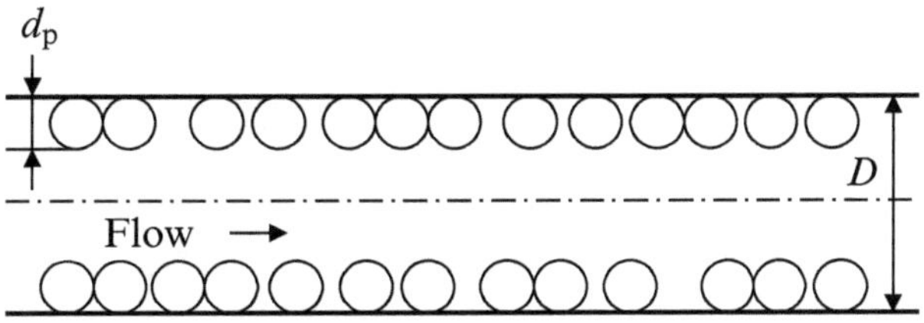

Fig. 2.17 Sand roughness in a circular pipe

2.4.4 *Friction Coefficient, λ*

It is of practical significance to estimate the pressure drop in circular and noncircular pipes. Expressions are introduced in this chapter for friction coefficients in the fully developed region of laminar and turbulent pipe flows.

2.4.4.1 Pipe Roughness

The inner wall of a circular pipe used over a prolonged time for flow delivery is known to be quite rough. The wall roughness affects the pressure drop significantly for turbulent pipe flows. It is therefore necessary to distinguish between smooth and rough pipes. The following dimensionless number is employed to judge whether the wall of a circular pipe is smooth or rough:

$$v * d_p/\nu_f = 70 \tag{2.77}$$

where

$$v* = (\tau_w/\rho)^{1/2} \tag{2.78}$$

is the friction velocity, d_p is the roughness height, ν_f is the kinematic viscosity of fluid, τ_w is the wall shear stress, and ρ is the density of fluid (see Fig. 2.17). The wall of a pipe is smooth for $v * d_p/n_f \leq 70$, while it is rough for $v * d_p/n_f > 70$.

2.4.4.2 Friction Coefficient in a Smooth Circular Pipe

Laminar Pipe Flow

When a flow in a circular pipe is laminar, the friction coefficient, λ, is given by

$$\lambda = 64/Re \tag{2.79}$$

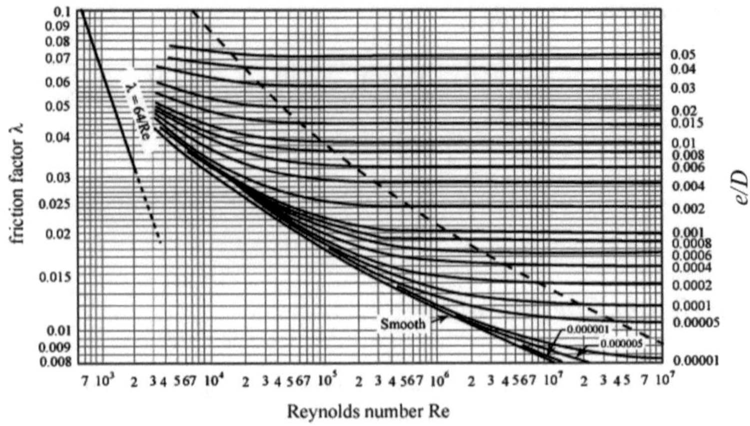

Fig. 2.18 Moody diagram

Turbulent Pipe Flow

Many empirical equations have been proposed for turbulent flow in a circular pipe. Some of these equations are listed below:

$$\text{Blasius} : \lambda = 0.3164 Re^{-1/4} \quad (2,320 < Re < 10^5) \tag{2.80}$$

$$\text{Nikuradse} : \lambda = 0.0032 + 0.221 Re^{-0.237} \; (10^5 < Re < 10^8) \tag{2.81}$$

$$\text{Prandtl-Karman-Nikuradse} : 1/\lambda = \left[2\log_{10}(Re\lambda^{1/2}) - 0.8\right]^2 \tag{2.82}$$

2.4.4.3 Friction Coefficient in a Rough Circular Pipe

Colebrook proposed the following empirical equation for turbulent flow in rough pipes:

$$1/\lambda = \left[2.0\log_{10}\{D/(2e)\} + 1.74\right]^2 \tag{2.83}$$

where D is the pipe diameter and e is the height of the roughness element. The friction coefficient, λ, in rough pipes is given in Fig. 2.18, known as the Moody diagram.

2.4.4.4 Friction Coefficient in a Rectangular Duct

The friction coefficient of laminar flow in a smooth rectangular duct (see Fig. 2.19) is expressed by

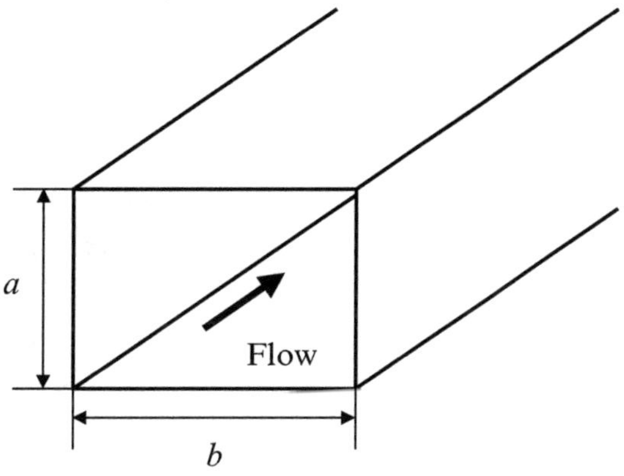

Fig. 2.19 Flow in a rectangular duct

$$\lambda Re = 96(1 - 1.3553a* + 1.9467\alpha*^2 - 1.7012\alpha*^3 + 0.9564\alpha*^4 - 0.2537\alpha*^5) \tag{2.84}$$

$$a* = a/b \tag{2.85}$$

2.4.4.5 Friction Coefficient for Laminar Flow in Smooth Concentric Annuli (Fig. 2.20)

$$\lambda Re = 64(1 - r*)^2/\left[1 + r*^2 - 2r_m*^2\right] \tag{2.86}$$

$$r* = d/D \tag{2.87}$$

$$r_m* = \left[(1 - r*^2)/\{2\ln(1/r*)\}\right]^{1/2} \tag{2.88}$$

$$Re = v_m D_h/\nu \tag{2.89}$$

$$D_h = D - d \tag{2.90}$$

where d and D denote the inner and outer diameters, respectively, and D_h is the hydraulic diameter defined in the next section.

Exercise 2.10 Silicone oil with kinematic viscosity $\nu_f = 100\ \mathrm{mm^2/s}$ flows in a smooth pipe of diameter $D = 10$ mm (see Fig. 2.21). The cross-sectional mean velocity, $v_{m,}$ is 20 mm/s and the critical Reynolds number, Re_c, is 2,320. Calculate the frictional coefficient, λ.

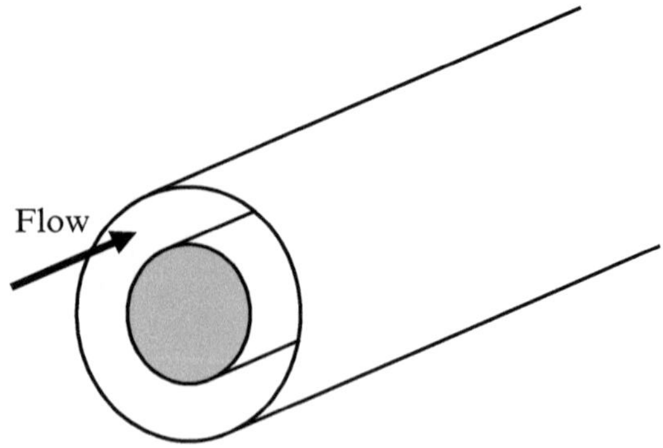

Fig. 2.20 Flow in a concentric annuli (outer and inner diameters are denoted by D and d, respectively)

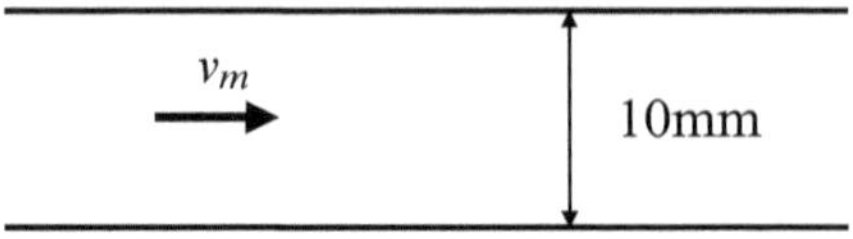

Fig. 2.21 Flow in a circular pipe

Answer to Exercise 2.10 The Reynolds number, Re, is given by

$$Re = v_m D/\nu_f = 20 \times 10^{-3} \times 10 \times 10^{-3}/(100 \times 10^{-6}) = 2.0 \quad (2.91)$$

Since the flow is laminar, the friction coefficient, λ, can be calculated from Eq. (2.79) as follows:

$$\lambda = 64/Re = 64/2.0 = 32 \quad (2.92)$$

Exercise 2.11 Water with kinematic viscosity $\nu_f = 0.892\ \mathrm{mm^2/s}$ flows in a smooth pipe of diameter $D = 50$ mm. The cross-sectional mean velocity, v_m, is 1.50 m/s. Calculate the friction coefficient, λ.

Answer to Exercise 2.11 The Reynolds number, Re, is given by

$$Re = v_m D/\nu_f = 1.50 \times 50 \times 10^{-3}/(0.892 \times 10^{-6}) = 8.41 \times 10^4 \quad (2.93)$$

The friction coefficient, λ, is calculated from the Blasius equation to give

$$\lambda = 0.3164 Re^{-1/4} = 0.3164(8.41 \times 10^4)^{-1/4} = 0.0186 \quad (2.94)$$

Exercise 2.12 Compare the friction coefficients calculated from the correlations proposed by Blasius, Nikuradse, and Prandtl–Karman–Nikuradse for Reynolds number, Re, of 1.0×10^5.

Answer to Exercise 2.12 Blasius:

$$\lambda = 0.3164 Re^{-1/4} = 0.3164 \times (1.0 \times 10^5)^{-1/4} = 0.0178 \tag{2.95}$$

Nikuradse:

$$\lambda = 0.0032 + 0.221 Re^{-0.237} = 0.0032 + 0.221 \times (1.0 \times 10^5)^{-0.237} = 0.0176 \tag{2.96}$$

Prandtl–Karman–Nikuradse:

The root of Eq. (2.82) can be numerically obtained to give

$$\lambda = 0.0180 \tag{2.97}$$

2.4.5 *Hydraulic Diameter,* $\mathbf{D}_h$

Compared to flows in a circular pipe, information on the friction coefficient in noncircular pipes is limited. In order to apply the friction coefficient expressions proposed originally for circular pipes to noncircular pipes, the pipe diameter is replaced by the following hydraulic diameter, D_h, defined thus:

$$D_\mathrm{h} = \frac{4A}{P} \tag{2.98}$$

where A and P are the cross-sectional area and the peripheral length (or wetted perimeter) of the noncircular pipe, respectively (see Fig. 2.22).

Exercise 2.13 Obtain the hydraulic diameter for a rectangular pipe shown in Fig. 2.23. The side lengths of the pipe are denoted by a and b, respectively.

Answer to Exercise 2.13 The cross-sectional area, A, and the peripheral length, P, are respectively given as follows:

$$A = ab \tag{2.99}$$

$$P = 2(a + b) \tag{2.100}$$

Substituting Eqs. (2.99) and (2.100) into Eq. (2.98) yields

$$D_\mathrm{h} = \frac{4A}{P} = \frac{4ab}{2(a+b)} = \frac{2ab}{a+b} \tag{2.101}$$

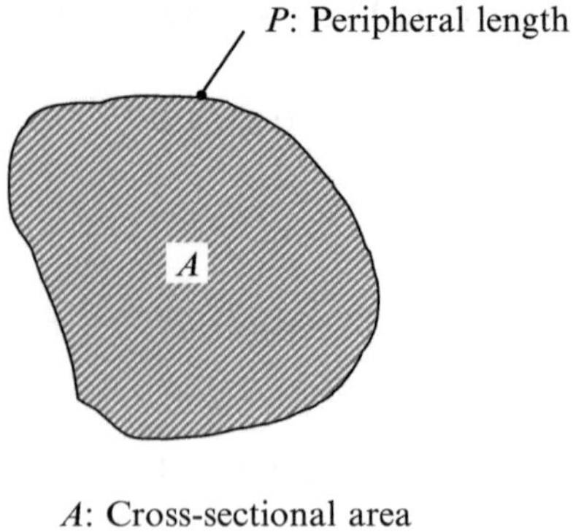

Fig. 2.22 Cross section of a non-circular duct

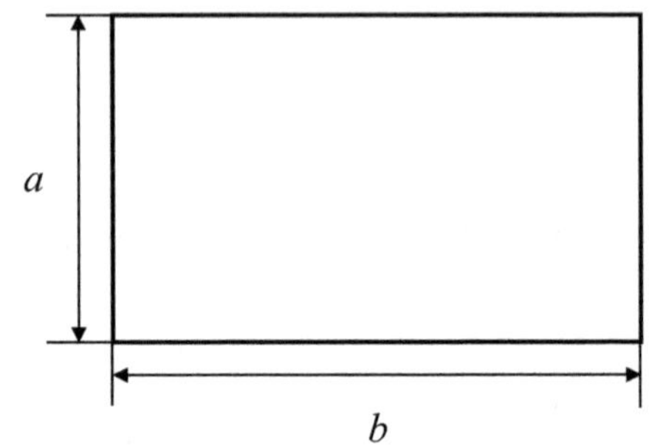

Fig. 2.23 Cross section of a rectangular duct

Exercise 2.14 Obtain the hydraulic diameter for a triangular pipe (see Fig. 2.24). The length of each side of the pipe is a.

Answer to Exercise 2.14

$$A = \frac{\sqrt{3}a^2}{4} \tag{2.102}$$

$$P = 3a \tag{2.103}$$

$$D_\mathrm{h} = \frac{4A}{P} = \frac{4\sqrt{3}a^2/4}{3a} = \frac{\sqrt{3}a}{3} \tag{2.104}$$

Exercise 2.15 Obtain the hydraulic diameter of a passage between two parallel plates. The clearance of the passage is a.

Answer to Exercise 2.15 Taking the limit of b ($b \to \infty$) in Eq. (2.101) gives $D_\mathrm{h} = 2a$.

Exercise 2.16 Verify that $D_\mathrm{h} = D$ for a circular pipe.

Answer to Exercise 2.16 Substituting $A = \pi D^2/4$ and $P = \pi D$ into Eq. (2.98) yields $D_\mathrm{h} = D$.

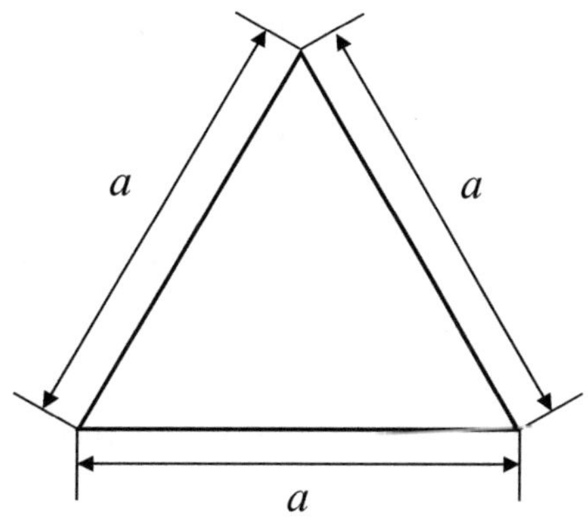

Fig. 2.24 Cross section of a triangular duct

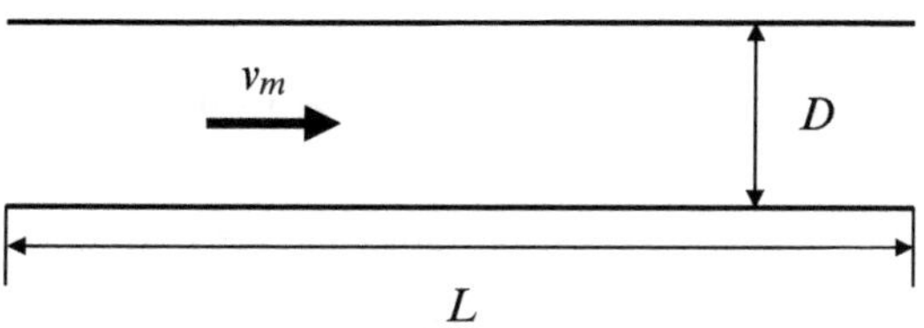

Fig. 2.25 Flow in a circular pipe (*D*: diameter, *L*: length)

Exercise 2.17 Water flows in a horizontal circular pipe of inner diameter, *D*, of 45 mm and length, *L*, of 250 m, as shown in Fig. 2.25. The temperature of the water is 20 °C and the flow rate, Q_L, is 1.23×10^{-3} m^3/s. Obtain the following quantities by considering that the hydrodynamic loss is caused only by friction:

(a) Cross-sectional mean velocity, $v_{m.}$
(b) Reynolds number, *Re*.
(c) Determine whether the flow is laminar or turbulent. The critical Reynolds number is 2,320.
(d) Frictional coefficient, λ (see Fig. 2.18).
(e) Pressure loss, Δp.

Answer to Exercise 2.17

(a) $$v_m = Q/A = 1.23 \times 10^{-3}/[\pi \times (0.045)^2/4] = 0.774 \text{ m/s} \quad (2.105)$$

(b) $$Re = v_m D/\nu_f = 0.774 \times 0.045/(1.00 \times 10^{-6}) = 3.48 \times 10^4 \quad (2.106)$$

(c) Turbulent since $Re > 2{,}320$

(d) $$\lambda = 0.3164 Re^{-1/4} = 0.3164(3.47 \times 10^4)^{-1/4} = 0.0232 \quad (2.107)$$

(e) $$\Delta p = \lambda(L/D)\rho {v_m}^2/2 = 0.0232 \times (250/0.045) \times 997 \times (0.774)^2/2 = 3.85 \times 10^4 \text{ Pa} \quad (2.108)$$

2.4.6 *Pipe Elements and Loss Coefficient, ζ*

2.4.6.1 Pipe Elements

There are many types of pipe elements in pipeline systems. The common elements in real pipeline systems are briefly described below:

Sudden expansion: The cross-sectional area of the pipe increases suddenly, as shown in Fig. 2.26.

Sudden contraction: The cross-sectional area of the pipe decreases suddenly (see Fig. 2.27).

Diffuser: The cross-sectional area of the pipe increases gradually in subsonic flow (see Fig. 2.28).

Nozzle: The cross-sectional area of the pipe decreases gradually in subsonic flow (see Fig. 2.29).

Bend: The pipe bends gradually (see Fig. 2.30).

Elbow: The pipe bends suddenly (see Fig. 2.31).

Orifice: The cross-sectional area of pipe is locally decreased using a hollow plate (see Fig. 2.32).

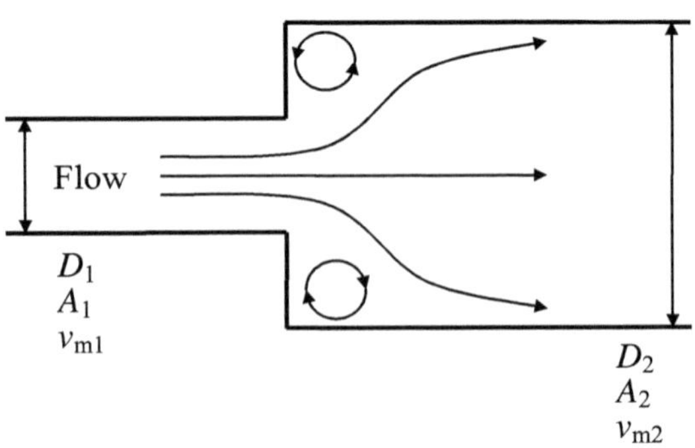

Fig. 2.26 Flow in a sudden expansion

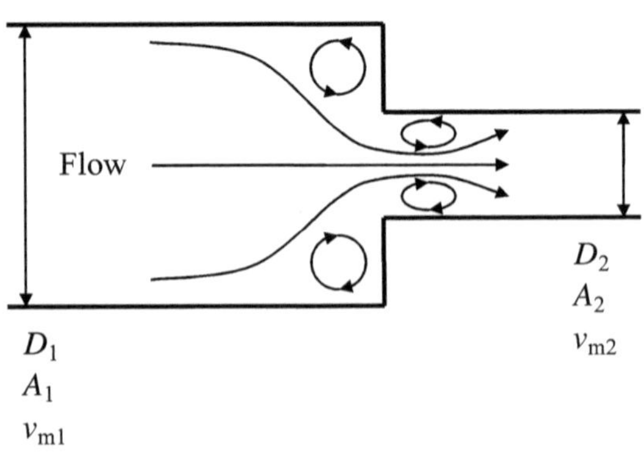

Fig. 2.27 Flow in a sudden contraction

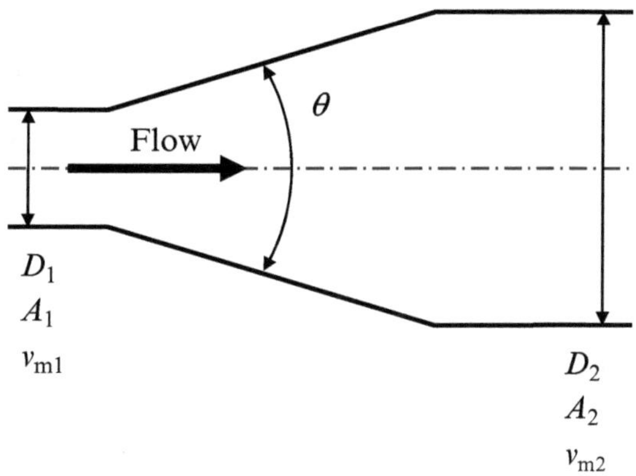

Fig. 2.28 Flow in a diffuser

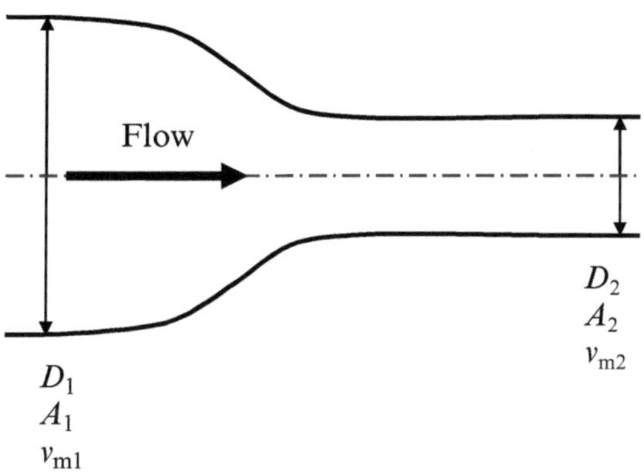

Fig. 2.29 Flow in a nozzle

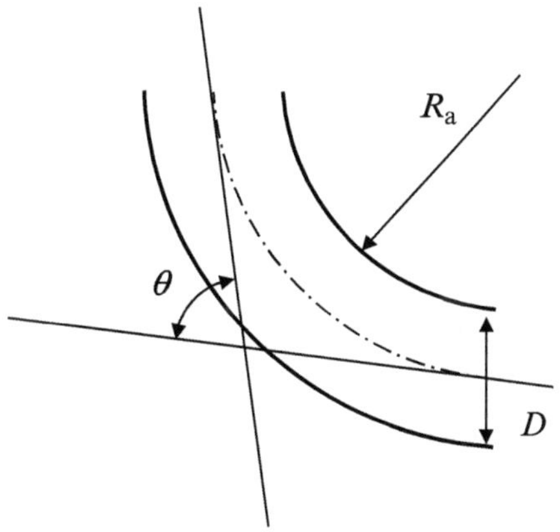

Fig. 2.30 Flow in a bend

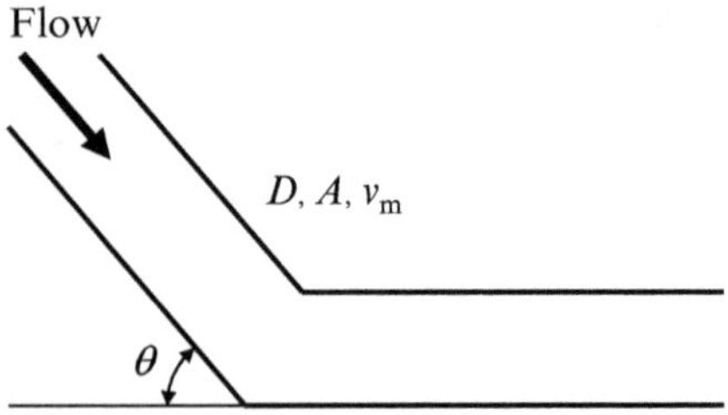

Fig. 2.31 Flow in an elbow

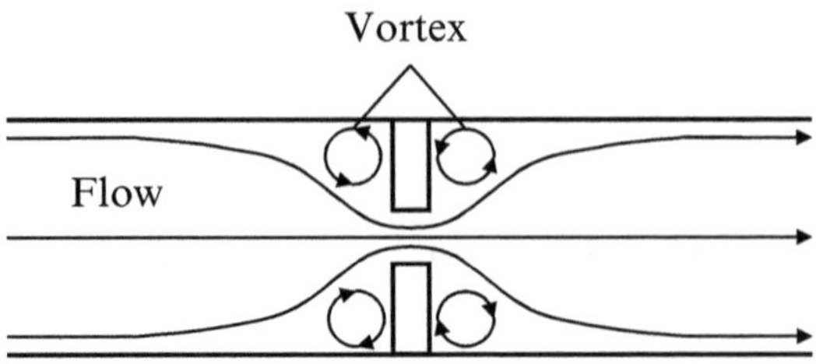

Fig. 2.32 Flow through a plate orifice

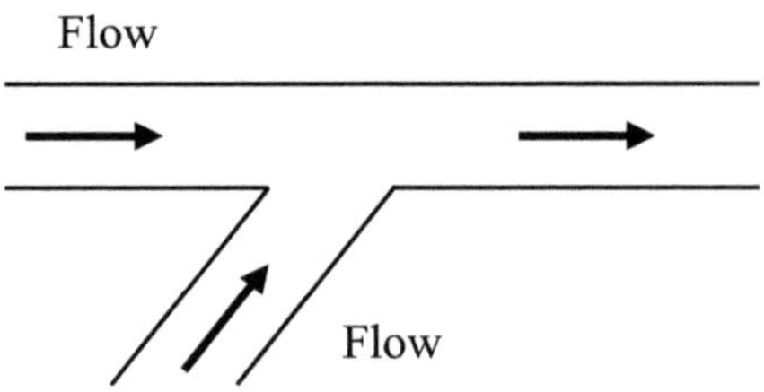

Fig. 2.33 Flow in a combining junction

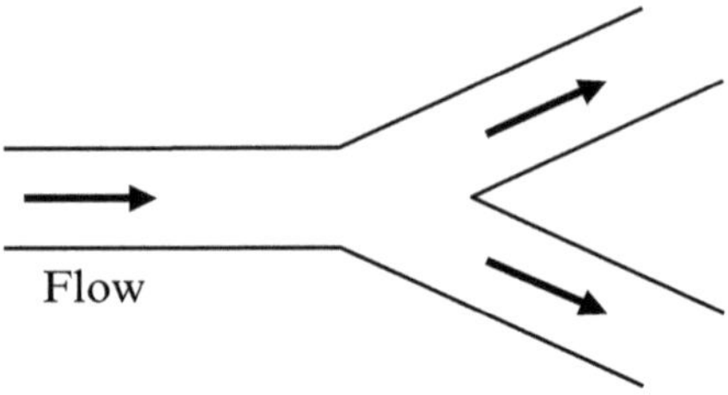

Fig. 2.34 Flow in a dividing junction

Combining junction: More than two pipes are connected to a single pipe (see Fig. 2.33).

Dividing junction: A single pipe is divided into plural pipes (see Fig. 2.34).

2.4.6.2 Definition of Loss Coefficient ζ

The loss coefficient, ζ, is defined in the following manner:

$$\Delta p = \rho g h_L \tag{2.109}$$

where Δp is the pressure drop, ρ is the density of fluid, g is the acceleration due to gravity, and h_L is the head loss.

$$h_L = \zeta v_m{}^2/(2g) \tag{2.110}$$

2.4.6.3 Loss Coefficient of Pipe Elements

Sudden Expansion

$$h_L = \zeta v_{m1}{}^2/(2g) \tag{2.111}$$

$$\zeta = \xi(1 - m)^2 \tag{2.112}$$

$$m = A_1/A_2 \tag{2.113}$$

where v_{m1} is the cross-sectional mean velocity in the smaller pipe upstream of the sudden expansion, m is the expansion ratio, ξ is the coefficient and usually assumed to be unity, and A_1 and A_2 are the cross-sectional areas of the upstream and downstream pipes, respectively.

Exercise 2.18 The smaller and larger diameters of a sudden expansion, D_1 and D_2, are 5.0 and 10.0 cm, respectively. Estimate the loss coefficient of the sudden expansion by assuming that the coefficient, ξ, in Eq. (2.112) is unity.

Answer to Exercise 2.18 The area ratio, m, is expressed by

$$m = A_1/A_2 = (D_1/D_2)^2 = (5.0/10.0)^2 = 1/4 = 0.25 \tag{2.114}$$

The loss coefficient is calculated from Eq. (2.113) to give

$$\zeta = \xi(1 - m)^2 = 1 \times (1 - 0.25)^2 = 0.5625 \tag{2.115}$$

Sudden Contraction

$$h_L = \zeta v_{m2}{}^2/(2g) \tag{2.116}$$

$$\zeta = (1/C_a - 1)^2 \tag{2.117}$$

where v_{m2} is the cross-sectional mean velocity in the downstream (smaller) pipe and C_a is the contraction coefficient. Merriman's equation is commonly used for evaluating C_a, thus [6]:

$$C_a = 0.582 + \frac{0.048}{1.1 - D_2/D_1} \tag{2.118}$$

Diffuser

$$h_L = \xi(v_{m1} - v_{m2})^2/(2g) \tag{2.119}$$

$$\xi = 3.50(\tan\theta/2)^{1.22} \tag{2.120}$$

where θ is the angle of expansion (°).

Nozzle

$$h_L = 0.04{v_{m2}}^2/(2g) \tag{2.121}$$

The loss coefficient of a nozzle is very small compared to that of sudden contraction. This means that the generation of vortex (or vortices) downstream of the corner (or corners) significantly affects the loss coefficient.

Bend

$$h_L = \zeta {v_m}^2/(2g) \tag{2.122}$$

$$\zeta = \left[0.131 + \frac{1.84}{[D/(2R_a)]^{3.5}}\right]\frac{\theta}{90} \tag{2.123}$$

where R_a is the radius of curvature of the bend and θ is the deflection angle (°), as shown in Fig. 2.30.

Elbow

$$h_L = \zeta {v_m}^2/(2g) \tag{2.124}$$

$$\zeta = 0.946\sin^2(\theta/2) + 2.05\sin^4(\theta/2) \tag{2.125}$$

where θ denotes the deflection angle (°), as shown in Fig. 2.31.

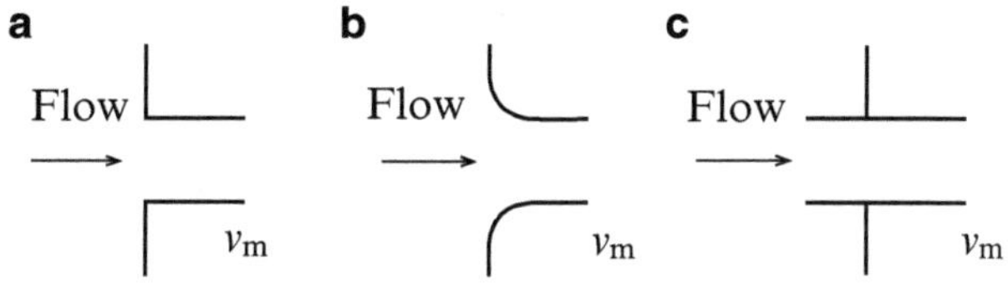

Fig. 2.35 Flow entering various types of entrances (**a**) Sharp corner (**b**) Round corner (**c**) Borda mouthpiece

Orifice

$$h_L = \zeta v_m^2/(2g) \tag{2.126}$$

$$\zeta = [1/(C_a m) - 1]^2 \tag{2.127}$$

$$m = A_o/A \tag{2.128}$$

where A_o is the cross-sectional area of the orifice plate, and A is the cross-sectional area of the pipe. The contraction coefficient, C_a, is given, for example, by JIS (Japanese Industrial Standard). An orifice is widely used as a flow meter. The accuracy is quite excellent in spite of the simple design.

Pipe Entrance

Three pipe entrance configurations that are typical are shown in Fig. 2.35. The loss coefficient values for configurations (a), (b), and (c) are approximately 0.50, 0.05, and 0.56, respectively. It should be noted that the loss coefficient of the pipe entrance depends strongly on the shape of the entrance. The loss coefficient is closely associated with the generation of vortices around the corners.

Pipe Exit

The velocity head of fluid issuing out of the pipe exit is regarded as an energy loss, since the fluid is expanding into a seemingly unbounded infinite medium,

$$h_L = v_m^2/(2g) \tag{2.129}$$

Combining Junction and Dividing Junction

Figure 2.33 shows that more than two pipes merge into a single pipe. This configuration is called the combining junction. On the other hand, Fig. 2.34 illustrates a single pipe dividing into more than two pipes. Such a structure is called

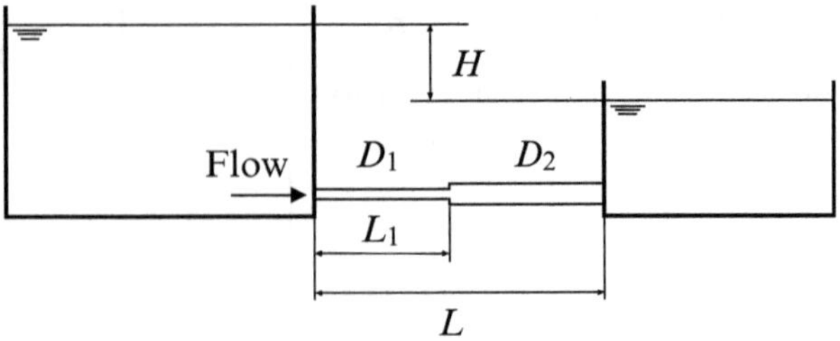

Fig. 2.36 Pipe line connecting two large reservoirs

the dividing junction. Vortices generated around the junction causes an energy loss. The details of the loss coefficients of the two types of junctions are available in the references listed at the end of this book.

Exercise 2.19 There is a horizontal, circular pipe connecting two big reservoirs, 1 and 2 (see Fig. 2.36). The length of the pipe is 100 m. The pipe diameter is suddenly increased from 300 to 600 mm at the midpoint between the two reservoirs. The loss coefficient at the pipe entrance, ζ_I, is 0.50. The frictional coefficients, λ_1 and λ_2, in the smaller and larger pipes are 0.030 and 0.020, respectively. The water flow rate, Q, is 18 m^3/min. The loss coefficient at sudden expansion, ζ_{se}, is expressed by

$$\zeta_{se} = (1 - m)^2 \tag{2.130}$$

$$m = A_1/A_2 \tag{2.131}$$

where m is the area ratio. Obtain the water surface height difference, H.

Answer to Exercise 2.19 The liquid velocities, v_{s1} and v_{s2}, on the two reservoir surfaces are negligibly small since the reservoirs are very large. The cross-sectional mean velocities in the smaller and larger pipes, v_{m1} and v_{m2}, are calculated as

$$v_{m1} = Q/[\pi d_1{}^2/4] = (18/60)/[3.14 \times (0.300)^2/4] = 4.24\ \mathrm{m/s} \tag{2.132}$$

$$v_{m2} = Q/[\pi d_2{}^2/4] = (18/60)/[3.14 \times (0.600)^2/4] = 1.06\ \mathrm{m/s} \tag{2.133}$$

The Bernoulli equation results in

$$\frac{p_{BS1}}{\rho g} + \frac{v_{mBS1}{}^2}{2g} + z_{Bs1} = \frac{p_{Bs2}}{\rho g} + \frac{v_{mBS2}{}^2}{2g} + z_{BS2} + h_L \tag{2.134}$$

where $p_{BS1} = p_{BS2} = p_0$, $v_{mBS1} = v_{mBS2} = 0$, $z_{BS1} - z_{BS2} = H$, and h_L is the hydrodynamic loss.

The water surface difference, H, therefore is

$$H = h_L \tag{2.135}$$

where h_L is expressed by

$$h_L = \zeta_i \frac{v_{m1}^2}{2g} + \lambda_1 \frac{l_1}{d_1} \frac{v_{m1}^2}{2g} + \zeta_{se} \frac{v_{m1}^2}{2g} + \lambda_2 \frac{l_2}{d_2} \frac{v_{m2}^2}{2g} + \zeta_e \frac{v_{m2}^2}{2g} \tag{2.136}$$

and ζ_{se} is calculated as

$$\zeta_{se} = (1 - A_1/A_2)^2 = \left[1 - (d_1/d_2)^2\right]^2 = (1 - 1/4)^2 = 0.5625 \tag{2.137}$$

Accordingly, H is given by

$$\begin{aligned} H = {} & 0.5 \times (4.24)^2/19.6 + 0.03 \times (50/0.300) \times (4.24)^2/19.6 \\ & + 0.5625 \times (4.24)^2/19.6 + 0.02 \times (50/0.600) \times (1.06)^2/19.6 \\ & + 1.0 \times (1.06)^2/19.6 = 5.72 \text{ m} \end{aligned} \tag{2.138}$$

2.5 Boundary Layer Flow on a Flat Plate

In 1904 Prandtl indicated that the flow about a solid body can be divided into two regions: a very thin layer close to the body (boundary layer) where friction is dominant and the remaining region outside this layer, where friction may be neglected [1].

2.5.1 Laminar Flow

Consider a flat plate placed in a fluid flowing at an approach velocity of V, as shown in Fig. 2.37. A laminar boundary layer initially develops on the plate. The transition from laminar flow to turbulent flow occurs at a certain distance from the leading edge of the plate. It is possible to induce transition to turbulence near the leading edge by placing a trap wire on the plate in the close vicinity of the leading edge (see Fig. 2.38).

The boundary layer thickness, δ, is defined as the distance from the plate to the position at which the axial velocity component, u, is 0.99 V (see Fig. 2.39). The thickness of laminar boundary layer is expressed by

$$\delta = 5.0(\nu_f x/V)^{1/2} = 5.0 x {Re_x}^{-1/2} \tag{2.139}$$

$$Re_x = Vx/\nu_f \tag{2.140}$$

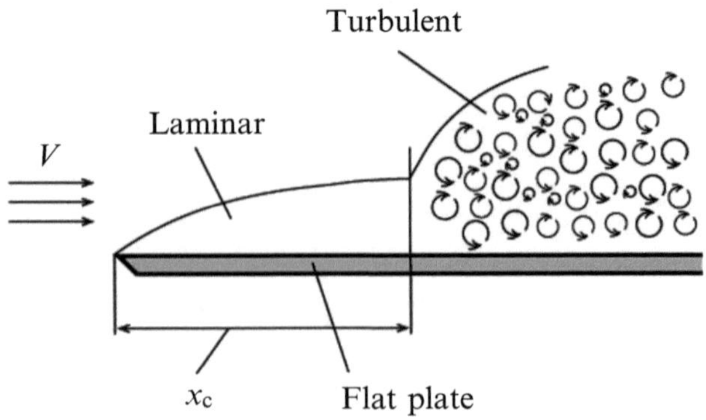

Fig. 2.37 Boundary layer on a flat plate

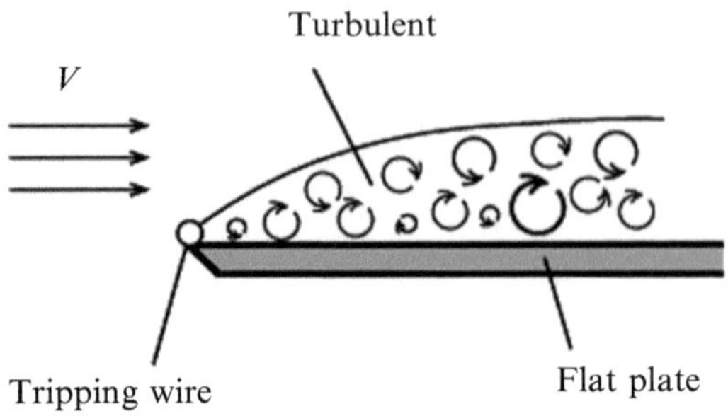

Fig. 2.38 Turbulent boundary layer caused by a tripping wire placed on the leading edge of a flat plate

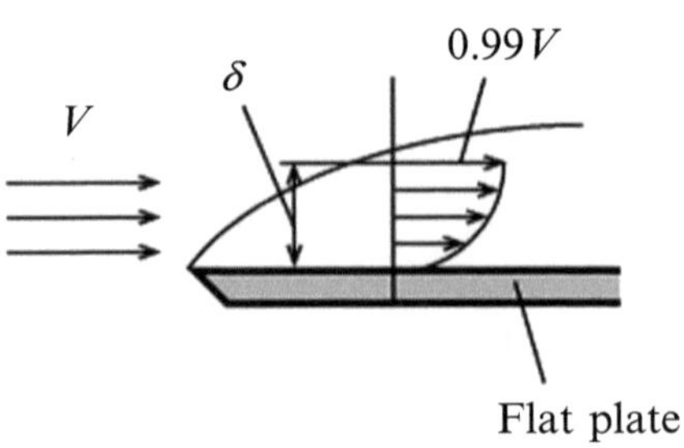

Fig. 2.39 Velocity distribution in a flat plate boundary layer and the definition of the boundary layer thickness, δ

where ν_f is the kinematic viscosity of fluid and x is the distance from the leading edge.

The velocity distribution in a laminar boundary layer is given by the Blasius solution [1]. The wall shear stress, τ_w, is expressed as

$$\tau_w = C_f \rho V^2/2 \tag{2.141}$$

$$C_f = 0.664 {Re_x}^{-1/2} \tag{2.142}$$

where C_f is the skin friction coefficient.

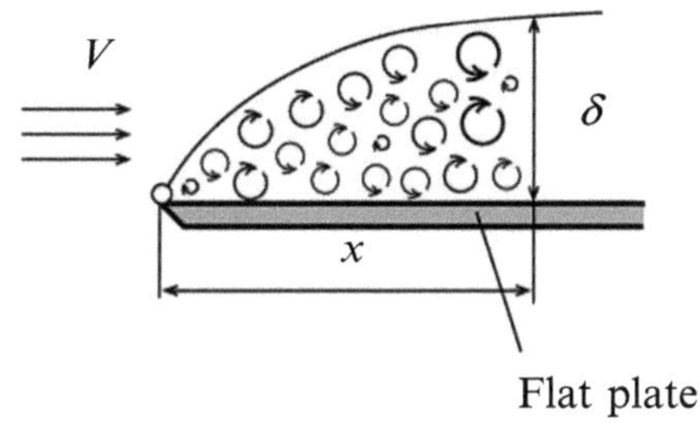

Fig. 2.40 Boundary layer thickness in a turbulent boundary layer

The mean skin friction coefficient per unit width is given by

$$C_{\mathrm{fm}} = 1.328 Re_L^{-1/2} \quad (2.143)$$

$$Re_L = VL/\nu_{\mathrm{f}} \quad (2.144)$$

Equation (2.143) is derived by averaging C_{f} between $x = 0$ and $x = L$. The fictional force acting on a flat plate can be obtained by using this skin friction coefficient.

2.5.2 Turbulent Flow

The following empirical relations have been proposed by assuming that the flow becomes turbulent from the leading edge of the plate. This situation is realized by placing a trap wire near the leading edge, as mentioned earlier (see Fig. 2.40).

The thickness of a turbulent boundary layer on a flat plate is

$$\delta = 0.37x(Vx/\nu_{\mathrm{f}})^{-1/5} \quad (2.145)$$

The local and mean friction coefficients are respectively given by

$$C_{\mathrm{f}} = (2.0 \log Re_x - 0.65)^{-2.3} \quad (2.146)$$

$$C_{\mathrm{fm}} = 0.455(\log Re_L)^{-2.58} \quad (2.147)$$

2.5.3 Transition Flow

The local Reynolds number, Re_x, is defined as

$$Re_x = Vx/\nu_{\mathrm{f}} \quad (2.148)$$

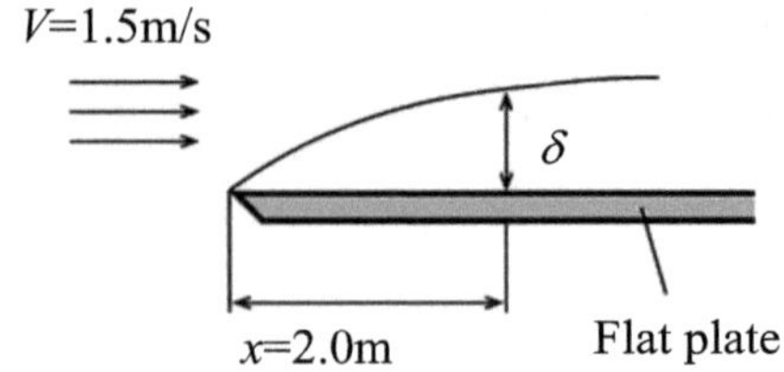

Fig. 2.41 Laminar flat plate boundary layer

The critical Reynolds number for transition to turbulence on a smooth flat plate boundary layer (see Fig. 2.37) is expressed by

$$Re_{xc} = Vx_c/\nu_f = 5 \times 10^5 \tag{2.149}$$

Exercise 2.20 The velocity of airflow approaching a flat plate, V, is 1.5 m/s (see Fig. 2.41). Calculate the local Reynolds number, Re_x, and the thickness of the boundary layer, δ, at the axial position $x = 2.0$ m. The kinematic viscosity of air, ν_f, is 15×10^{-6} m^2/s.

Answer to Exercise 2.20 The local Reynolds number, Re_x, is calculated as follows:

$$Re_x = Vx/\nu_f = 1.5 \times 2.0/(15 \times 10^{-6}) = 2.0 \times 10^5 \tag{2.150}$$

This local Reynolds number is smaller than Re_{xc} and, hence, the boundary layer is laminar. The boundary layer thickness, δ, is calculated from Eq. (2.139).

$$\begin{aligned} \delta = 5.0(\nu_f x/V)^{1/2} &= 5.0 \times (15 \times 10^{-6} \times 2.0/1.5)^{1/2} \\ &= 2.24 \times 10^{-2}\ \text{m} = 22.4\ \text{mm} \end{aligned} \tag{2.151}$$

Exercise 2.21 The velocity of airflow approaching a flat plate, V, is 1.5 m/s. Calculate the local Reynolds number, Re_x, and the thickness of the boundary layer, δ, at the axial position of $x = 6.0$ m. The kinematic viscosity of air, ν_f, is 15×10^{-6} m^2/s.

Answer to Exercise 2.21 The local Reynolds number, Re_x, is calculated as follows:

$$Re_x = Vx/\nu_f = 1.5 \times 6.0/(15 \times 10^{-6}) = 6.0 \times 10^5 \tag{2.152}$$

This local Reynolds number is greater than Re_{xc} and, hence, the boundary layer is turbulent there. The boundary layer thickness, δ, is given by

$$\delta = 0.37x(Vx/\nu_{\rm f})^{-1/5} = 0.37 \times 6.0 \times [1.5 \times 6.0/(15 \times 10^{-6}]^{-1/5}$$
$$= 0.155 \text{ m} = 155 \text{ mm} \tag{2.153}$$

Exercise 2.22 The velocity of airflow approaching a flat plate is 2.0 m/s. Calculate the local skin friction coefficient at $x = 1.5$ m and the mean skin friction coefficient between $x = 0$ and 1.5 m.

Answer to Exercise 2.22 The local Reynolds number is given by

$$Re_x = Vx/\nu_{\rm f} = 2.0 \times 1.5/(15 \times 10^{-6}) = 2.0 \times 10^5 \tag{2.154}$$

Since this Reynolds number is smaller than the critical Reynolds number, Re_{xc}, of 5×10^5, the two coefficients should be calculated from Eqs. (2.142) and (2.143), respectively.

$$C_{\rm f} = 0.664 {Re_x}^{-1/2} = 0.664 \times (2.0 \times 10^5)^{-1/2} = 1.48 \times 10^{-3} \tag{2.155}$$

$$Re_L = VL/\nu_{\rm f} = 2.0 \times 1.5/(15 \times 10^{-6}) = 2.0 \times 10^5 \tag{2.156}$$

$$C_{\rm fm} = 1.328 {Re_L}^{-1/2} = 1.328 \times (2.0 \times 10^5)^{-1/2} = 2.97 \times 10^{-3} \tag{2.157}$$

2.6 Flow Around a Solid Body

2.6.1 Flow Around a Sphere

Figure 2.42 shows a flow approaching a sphere. If the velocity of the flow is very small and, hence, the Reynolds number, Re, is smaller than unity ($Re < 1$), the fluid flows slowly along the sphere with no flow separation. Such a very slow flow is called creeping flow. When the Reynolds number exceeds a certain critical value, separation occurs and a reverse flow region develops around the rear part of the sphere. With further increase in V, vortices are formed and shed nearly periodically from the rear part of the sphere and, as a result, the wake behind the sphere becomes turbulent.

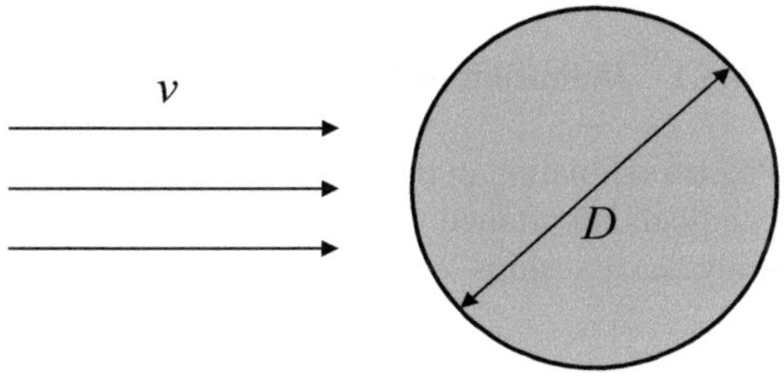

Fig. 2.42 Uniform flow approaching a sphere

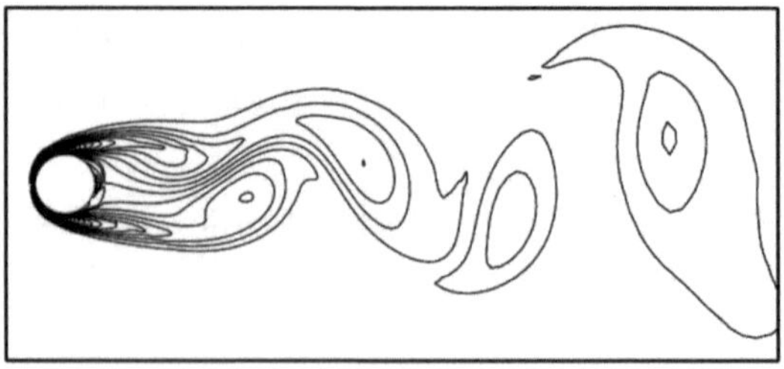

Fig. 2.43 Karman vortex streets formed behind a circular cylinder

The boundary layer on the sphere remains laminar even in the presence of the vortex shedding. When the Reynolds number attains another critical value, the boundary layer just upstream of the separation point undergoes transition to turbulence. This critical Reynolds number is approximately 3×10^5. With a further increase in the Reynolds number the turbulent boundary layer progressively extends towards the front stagnation point.

2.6.2 *Flow Around a Circular Cylinder*

The relationship between the flow pattern around a circular cylinder placed normal to the flow and the Reynolds number is basically the same as that around a sphere. The vortices successively shedding from the cylinder are specifically called Karman's vortices (see Fig. 2.43).

Exercise 2.23 The velocity, V, of water flow approaching a sphere 5.0 cm in diameter, D, is 0.15 m/s. Calculate the Reynolds number, Re. The kinematic viscosity of water, ν_f, is 1.00×10^{-6} m^2/s.

Answer to Exercise 2.23 The Reynolds number is

$$Re = VD/\nu_f = 0.15 \times 0.050/(1.00 \times 10^{-6}) = 7.5 \times 10^3 \qquad (2.158)$$

2.6.3 *Pressure Drag and Drag Coefficient*

2.6.3.1 Definition of Drag Coefficient, C_D

Precise estimation of drag force acting on a solid body immersed in a fluid flow is of practical importance for design of bodies in external flow including buildings, bridges, cars, ships, and airplanes. The drag force is expressed thus:

$$F_D = C_D A_p \rho V^2/2 \qquad (2.159)$$

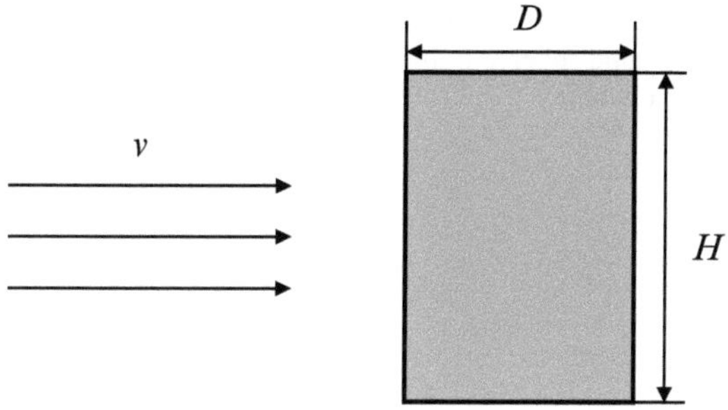

Fig. 2.44 Uniform flow approaching a cylindrical rod

where C_D is the drag coefficient, A_p is the area projected onto a plane normal to the flow, ρ is the density of fluid, and V is the approaching flow velocity. The expressions for A_p and C_D are (Fig. 2.44)

$$A_p = DH(\text{cylindrical rod, Fig. 2.44}) \tag{2.160}$$

$$= \pi D^2/4 \text{ (sphere)} \tag{2.161}$$

$$C_D = f(Re, Tu, \lambda_s/D) \tag{2.162}$$

where

$$Tu = v\prime_{rms}/V \tag{2.163}$$

In the above relations, Re is the Reynolds number, Tu is the turbulence intensity, λ_s/D is the relative scale of mean turbulent eddy, and v'_{rms} is the root-mean-square value of the turbulence component. The effect of λ_s/D is usually very small and the drag coefficient can be expressed by

$$C_D = f(Re, Tu) \tag{2.164}$$

When the turbulence intensity, Tu, is less than about 0.25 %, C_D is regarded as a sole function of Reynolds number, Re, as shown in Fig. 2.45, thus.

$$C_D = f(Re) \tag{2.165}$$

2.6.3.2 Drag Coefficient of a Sphere Immersed in a Flow of Very Low Turbulence Intensity

$$C_D = 24/Re \text{ (Stokes' resistance law}, Re < 1) \tag{2.166}$$

Fig. 2.45 Drag coefficients for a solid sphere, droplet, and a bubble

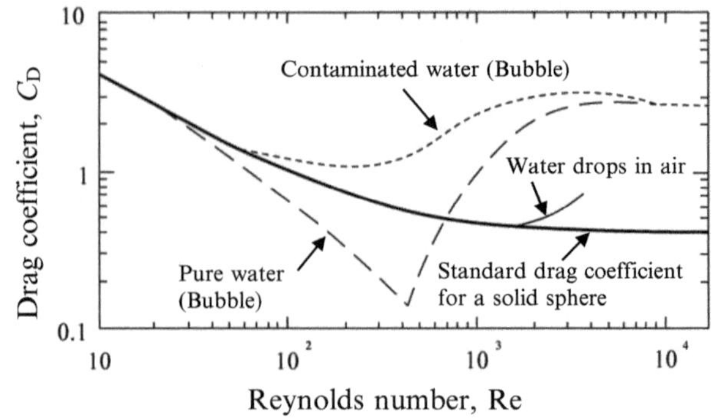

$$C_D = 10Re^{-1/2} \text{ (Allen's resistance law, } Re = 30 \sim 300) \tag{2.167}$$

$$C_D = 0.44 \text{ (Newton's resistance law, } Re = 300 \sim 10^5) \tag{2.168}$$

2.6.3.3 Drag Coefficient of Cylindrical Rod Placed Normal to a Flow of Very Low Turbulence Intensity

$$C_D = 8\pi/[Re(2.002 - \ln Re)] \quad (Re < 0.5) \tag{2.169}$$

$$C_D = (0.707 + 3.42Re^{-1/2})^2 \ (Re = 5 \sim 40) \tag{2.170}$$

$$C_D = 1.2 \ (Re = 2 \times 10^4 - 2 \times 10^5) \tag{2.171}$$

Exercise 2.24 A sphere 6.0 cm in diameter, D, is fixed in airflow (see Fig. 2.46). The approach velocity, V, is 4.0 m/s and the kinematic viscosity of air, ν_f, is 15×10^{-6} m^2/s. Calculate the Reynolds number, Re, and then estimate the drag coefficient, C_D.

Answer to Exercise 2.24 The Reynolds number, Re, is given by

$$Re = VD/\nu_f = 4.0 \times 6.0 \times 10^{-2}/(15 \times 10^{-6}) = 1.6 \times 10^4 \tag{2.172}$$

This Reynolds number falls in the regime where Newton's resistance law holds. The drag coefficient therefore is estimated from Eq. (2.168) to be 0.44.

Exercise 2.25 A sphere of diameter, D_p, of 8.0 cm and density, ρ_p, of 1505 kg/m^3 falls in still water at the terminal velocity (see Fig. 2.47). Obtain the weight, F_w, of the sphere, the buoyancy force acting on the sphere, F_B, and the terminal velocity, v_∞. The density of water, ρ_w, is 998 kg/m^3 and the drag coefficient of the sphere, C_D, is 0.44.

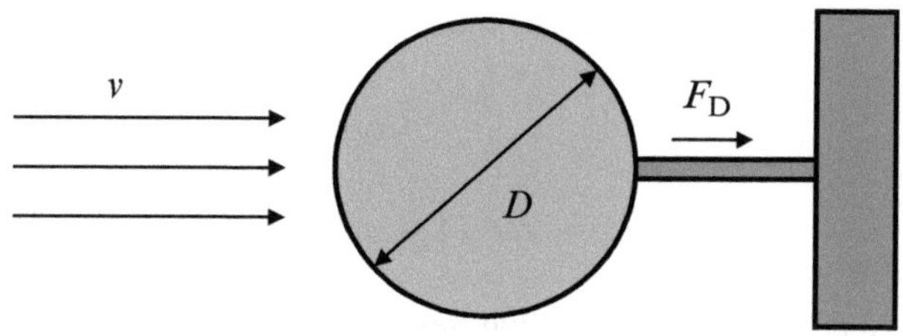

Fig. 2.46 Drag force acting on a sphere fixed in a uniform approaching flow

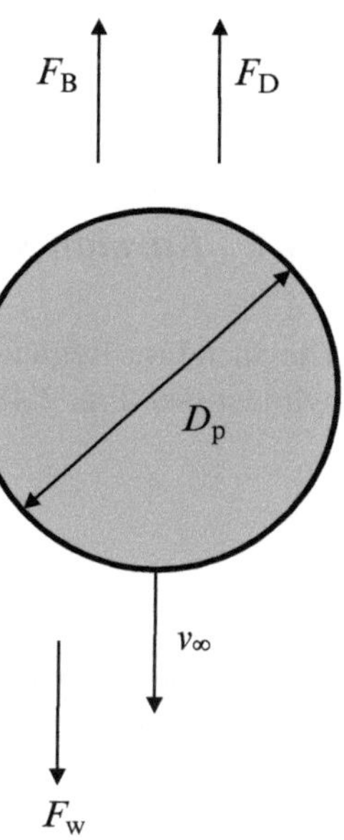

Fig. 2.47 Sphere falling in sill water at a terminal velocity

Answer to Exercise 2.25

$$F_w = \rho_p g \pi D_p{}^3/6 = 1505 \times 9.80 \times 3.14 \times (0.080)^3/6 = 3.95 \text{ N} \quad (2.173)$$

$$F_B = \rho_w g \pi D_p{}^3/6 = 998 \times 9.80 \times 3.14 \times (0.080)^3/6 = 2.63 \text{ N} \quad (2.174)$$

The hydrodynamic drag, F_D, is expressed by

$$F_D = C_D A_p \rho_w v_\infty{}^3/2 \quad (2.175)$$

The terminal velocity is attained when there is no net force acting on the body in the vertical direction and it falls at a constant steady (terminal) velocity. The force balance on the body is given by

$$F_D = F_w - F_B \quad (2.176)$$

Substituting Eq. (2.175) into Eq. (2.176) yields

$$C_D A_p \rho_w {v_\infty}^2/2 = 1.32 \tag{2.177}$$

$$A_p = \pi D^2/4 = \pi \times (0.080)^2/4 = 5.02 \times 10^{-3}\ \text{m}^2 \tag{2.178}$$

Accordingly, the terminal velocity is given by

$$v_\infty = [1.32 \times 2/(0.44 \times 5.02 \times 10^{-3} \times 998)^{1/2} = 1.10\ \text{m/s} \tag{2.179}$$

2.6.4 Karman's Vortex Street

The shedding frequency, *f*, of Karman's vortex streets formed behind a circular cylinder (see Fig. 2.48) can be correlated in terms of the Strouhal number, *St*.

$$St \equiv fD/V = 0.2\ (10^2 < Re < 10^5) \tag{2.180}$$

$$Re = VD/\nu_f \tag{2.181}$$

Equation (2.180) indicates that the Strouhal number, *St*, is constant over a wide Reynolds number range and that the shedding frequency, *f*, is proportional to the approaching flow velocity, *V*. This result is very useful for the measurement of fluid flow velocity and flow rate. Specifically, the velocity, *V*, can be determined by measuring the shedding frequency, *f*. The so-called vortex flow meter works on the basis of this principle. The sound induced by electric wires in a windy day is associated with the shedding of Karman's vortex streets from the wires.

Exercise 2.26 An electric wire 1.0 cm in diameter, *D*, is placed in the atmosphere. The wind velocity, *V*, and the kinematic viscosity of air, ν_a, are 30 m/s and 15×10^{-6} m^2/s, respectively. Calculate the shedding frequency of Karman's vortex streets, *f*.

Answer to Exercise 2.26 The Reynolds number, *Re*, is given by

$$Re = VD/\nu_f = 30 \times 1.0 \times 10^{-2}/(15 \times 10^{-6}) = 2.0 \times 10^4 \tag{2.182}$$

This Reynolds number falls in the regime where Karman's vortex streets are formed behind the wire. The shedding frequency is given by

$$f = 0.2V/D = 0.2 \times 30/0.010 = 600\ \text{Hz} \tag{2.183}$$

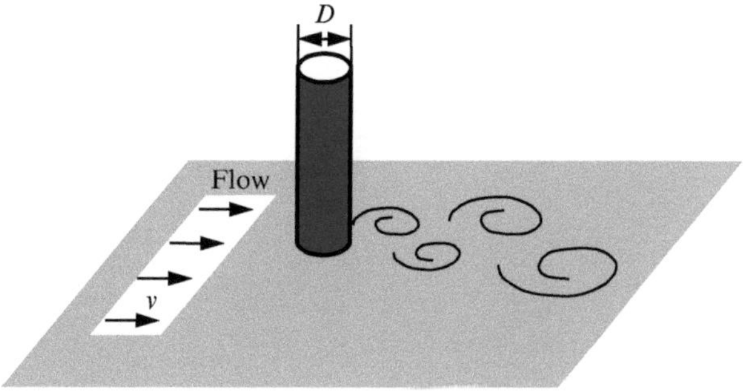

Fig. 2.48 Karman's vortex streets formed behind a circular cylinder placed normal to the flow direction

The Strouhal number, St, has a close relationship with the drag coefficient, C_D, over a wide Reynolds number range.

$$St = 0.21/{C_D}^{0.622} \ (10^3 < Re < 10^6) \tag{2.184}$$

2.7 Dimensional Analysis

Engineering equations must be dimensionally homogeneous. This implies that the units of all terms of any equation must be equal to one another. This fundamental principle is useful for deriving a functional relationship among parameters governing fluid flow phenomena. Two types of methods are used for dimensional analysis: the Lord Rayleigh method and the Buckingham Π theorem. The Lord Rayleigh method is valid for flow fields governed by parameters less than or equal to four. The Buckingham Π theorem is applicable when parameters are greater than or equal to five. The concepts of these methods will be introduced below with some examples.

2.7.1 Lord Rayleigh Method

This method is useful for flow phenomena governed by quantities less than or equal to four. One of the quantities is expressed as a product of the power of the other quantities.

We consider a pendant as shown in Fig. 2.49, consisting of a small sphere supported by a string. The period, t_p, is assumed to be affected by the length of the string, L, and the acceleration due to gravity, g. The relationship among these quantities is assumed as follows:

$$t_p = kL^{\alpha}g^{\beta} \tag{2.185}$$

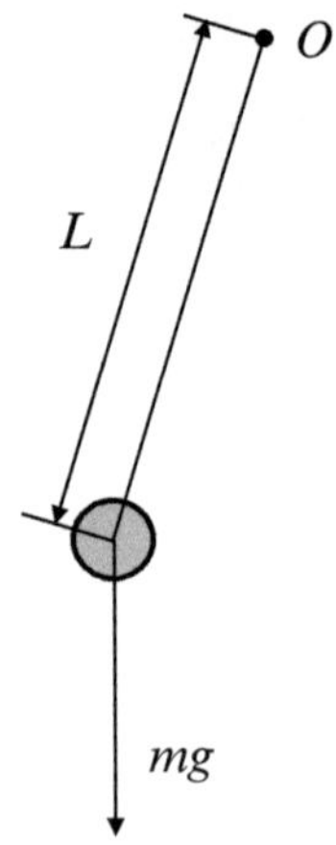

Fig. 2.49 Pendant moving under terrestrial condition

The two indices, α and β, are constant. The dimensions of T, L, and g are indicated by $[T]$, $[L]$, and $[LT^{-2}]$, respectively, in which $[T]$ represents dimension of time and $[L]$ for length.

The dimensions of the left-hand side and right-hand side of Eq. (2.185) must be equal to each other, thus:

$$[T] = [L]^{\alpha}[LT^{-2}]^{\beta} \tag{2.186}$$

This equation is rewritten as

$$[L^0T^1] = [L^{\alpha+\beta}T^{-2\beta}] \tag{2.187}$$

Using the law of exponents, the following two algebraic equations must be satisfied:

$$\alpha + \beta = 0 \tag{2.188}$$

$$-2\beta = 1 \tag{2.189}$$

Solving these equations yields

$$\alpha = 1/2 \tag{2.190}$$

$$\beta = -1/2 \tag{2.191}$$

Substituting these values into Eq. (2.185) gives

$$t_{\rm p} = k\left(\frac{L}{g}\right)^{1/2} \tag{2.192}$$

The exact solution is known to be

$$t_p = 2\pi \left(\frac{L}{g}\right)^{1/2} \tag{2.193}$$

It is evident that the Lord Rayleigh method leads to an exact relationship between the three quantities (T, L, and g).

Exercise 2.27 The propagation speed of sound, c, is assumed to be related to the pressure, p, density, ρ, and dynamic viscosity of fluid, μ. Derive the relationship between these quantities using the Lord Rayleigh method.

Answer to Exercise 2.27 The units of the four quantities are expressed as follows: c (m/s), p (kg/m s^2), ρ (kg/m^3), and μ (kg/m s). The speed of sound, c, is assumed to be

$$c = kp^{\alpha}\rho^{\beta}\mu^{\gamma} \tag{2.194}$$

where k is a constant. The relationship for the dimensions of Eq. (2.194) is expressed by

$$[L^1T^{-1}] = [ML^{-1}T^{-2}]^{\alpha}[ML^{-3}]^{\beta}[ML^{-1}T^{-1}]^{\gamma} \tag{2.195}$$

where L, T, and M denote the dimensions of length, time, and mass, respectively. The three indices, α, β, and γ, must satisfy the following algebraic equations:

$$L : -\alpha - 3\beta - \gamma = 1 \tag{2.196}$$

$$T : -2\alpha - \gamma = -1 \tag{2.197}$$

$$M : \alpha + \beta + \gamma = 0 \tag{2.198}$$

Solving these algebraic equations gives

$$\alpha = \frac{1}{2},\ \beta = -\frac{1}{2},\ \gamma = 0 \tag{2.199}$$

Substituting these values into Eq. (2.194) yields

$$c = k\left(\frac{p}{\rho}\right)^{1/2} \tag{2.200}$$

2.7.2 Buckingham Π Theorem

This theorem is applicable for a phenomenon governed by more than five quantities. When the number of quantities is n and number of dimensions is m, the following relationship holds among k (=$n - m$) dimensionless quantities:

$$f(\Pi_1, \Pi_2, \ldots \Pi_k) = 0 \tag{2.201}$$

or

$$\Pi_1 = f(\Pi_2, \ldots \Pi_k) = 0 \tag{2.202}$$

where Π_1, Π_2, ..., and Π_k are the dimensionless parameters. Unfortunately, it is difficult to determine the functional relationship based on this theorem.

Each dimensionless parameter is determined as a product of (m + 1) quantities among n quantities, where any one of n quantities must be used at least once. The following exercise will help the reader understand the process of obtaining the dimensionless parameters.

Exercise 2.28 A sphere is fixed in a uniform flow, as shown in Fig. 2.50. The approaching flow velocity, the diameter of the sphere, the density of fluid, and the dynamic viscosity of fluid are denoted by V (m/s), D (m), ρ (kg/m^3), and μ (kg/m s), respectively. Derive a relationship between the hydrodynamic drag, F_D (N = kgm/s^2), and these parameters based on the Buckingham Π theorem.

Answer to Exercise 2.28 The number of quantities, n, is 5 and number of dimensions, m, is 3. Thus there are two possible dimensionless parameters, Π_1 and Π_2. We choose V (m/s), D (m), and ρ (kg/m^3) as basic quantities to express

$$\Pi_1 = F_D V^{x1} D^{y1} \rho^{z1} \tag{2.203}$$

$$\Pi_2 = \mu V^{x2} D^{y2} \rho^{z2} \tag{2.204}$$

The following relationship holds concerning the dimensions of Eq. (2.203):

$$[L^0 T^0 M^0] = [MLT^{-2}][LT^{-1}]^{x1}[L]^{y1}[ML^{-3}]^{z1} \tag{2.205}$$

As the dimensions on both the sides must be the same, we have

$$L : 1 + x_1 + y_1 - 3z_1 = 0 \tag{2.206}$$

$$T : -2 - x_1 = 0 \tag{2.207}$$

$$M : 1 + z_1 = 0 \tag{2.208}$$

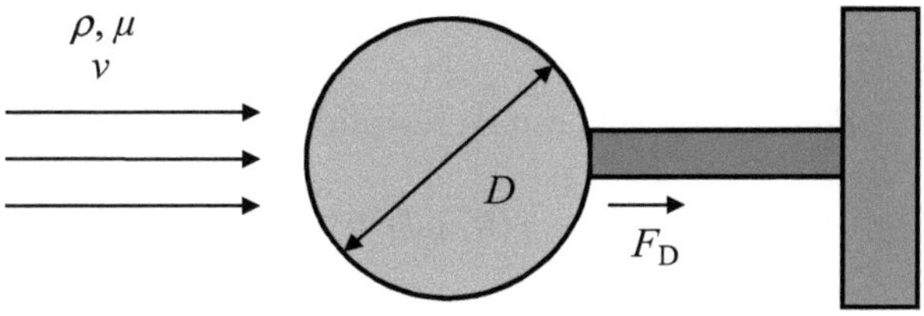

Fig. 2.50 A sphere fixed in a uniform approaching flow (*v*: approaching flow velocity, *D*: diameter, *ρ*: density, *μ*: dynamic viscosity)

Solving these algebraic equations yields

$$x_1 = -2, y_1 = -2, z_1 = -1 \tag{2.209}$$

Substituting these values into Eq. (2.203) gives

$$\Pi_1 = \frac{F_D}{V^2 D^2 \rho} \tag{2.210}$$

In a similar manner to the above derivation, the following results can be obtained for Eq. (2.204):

$$[L^0 T^0 M^0] = [ML^{-1}T^{-1}][LT^{-1}]^{x2}[L]^{y2}[ML^{-3}]^{z2} \tag{2.211}$$

$$L: -1 + x_2 + y_2 - 3z_2 = 0 \tag{2.212}$$

$$T: -1 - x_2 = 0 \tag{2.213}$$

$$M: 1 + z_2 = 0 \tag{2.214}$$

Solving these algebraic equations yields

$$x_2 = -1, y_2 = -1, z_2 = -1 \tag{2.215}$$

Substituting these values into Eq. (2.204) gives

$$\Pi_2 = \frac{\mu}{VD\rho} \tag{2.216}$$

where $\mu/(VD\rho)$ is the inverse of the Reynolds number, *Re*. Accordingly,

$$\Pi_1 = f(\Pi_2) = f(Re) \tag{2.217}$$

Combining Eq. (2.210) and Eq. (2.216) yields

$$F_{\mathrm{D}} = f(Re)\rho D^2 V^2 \tag{2.218}$$

Equation (2.218) can be rewritten as

$$\begin{aligned} F_{\mathrm{D}} = f(Re)\rho D^2 V^2 &= [8f(Re)/\pi](\pi D^2/4)(\rho V^2/2) \\ &= C_{\mathrm{D}} A_{\mathrm{p}} \rho V^2/2 \end{aligned} \tag{2.219}$$

where C_{D} is the drag coefficient and A_{p} is the area of the sphere projected to a plane perpendicular to the flow direction.

$$C_{\mathrm{D}} = [8f(Re)/\pi] \tag{2.220}$$

$$A_{\mathrm{p}} = \pi D^2/4 \tag{2.221}$$

Equation (2.219) is the well-known definition of the drag coefficient.

Exercise 2.29 The height of a liquid column lifted up in a capillary tube, H (m), is dependent on tube diameter, D (m), the density of liquid, ρ (kg/m^3), acceleration due to gravity, g (m/s^2), and the surface tension of liquid, σ (N/m = kg/s^2) (see Fig. 2.51). Express H as a function of D, ρ, g, and σ on the basis of the Buckingham Π theorem.

Answer to Exercise 2.29 The number of quantities, n, is 5 and number of dimensions, m, is 3 (L, T, M). There are two dimensionless parameters, Π_1 and Π_2. We choose D (m/s), ρ (kg/m^3), and g (m/s^2) as basic quantities to express

$$\Pi_1 = H D^{x1} \rho^{y1} g^{z1} \tag{2.222}$$

$$\Pi_2 = \sigma D^{x2} \rho^{y2} g^{z2} \tag{2.223}$$

The following relationship holds concerning the dimensions of Eq. (2.222):

$$[L^0 T^0 M^0] = [L][L]^{x1}[ML^{-3}]^{y1}[LT^{-2}]^{z1} \tag{2.224}$$

Since the dimensions on both the sides must be the same, we have

$$L : 1 + x_1 - 3y_1 + z_1 = 0 \tag{2.225}$$

$$T : -2z_1 = 0 \tag{2.226}$$

$$M : y_1 = 0 \tag{2.227}$$

Solving these algebraic equations yields

$$x_1 = -1, y_1 = 0, z_1 = 0 \tag{2.228}$$

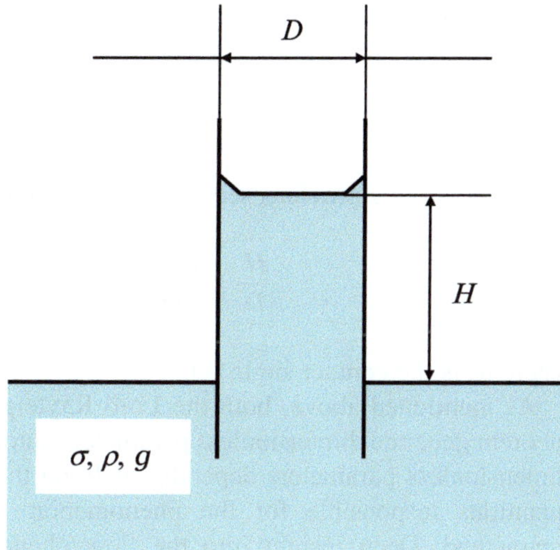

Fig. 2.51 Height of a liquid column lifted up in a capillary tube

Substituting these values into Eq. (2.222) gives

$$\Pi_1 = \frac{H}{D} \tag{2.229}$$

In a similar manner to the above derivation, the following results can be obtained for Eq. (2.223):

$$[L^0T^0M^0] = [MT^{-2}][L]^{x2}[ML^{-3}]^{y2}[LT^{-2}]^{z2} \tag{2.230}$$

$$L : x_2 - 3y_2 + z_2 = 0 \tag{2.231}$$

$$T : -2 - 2z_2 = 0 \tag{2.232}$$

$$M : 1 + y_2 = 0 \tag{2.233}$$

The solution of these algebraic equations is given as

$$x_2 = -2, y_2 = -1, z_2 = -1 \tag{2.234}$$

Substituting these values into Eq. (2.223) gives

$$\Pi_2 = \frac{\sigma}{D^2\rho g} \tag{2.235}$$

where $\sigma/(D^2\rho g)$ is a form of Bond number. Accordingly,

$$\frac{H}{D}=f\left(\frac{\sigma}{D^2\rho g}\right) \tag{2.236}$$

As shown in a previous section, the exact relationship is given by

$$\frac{H}{D}=\frac{4\sigma\cos\theta_c}{D^2\rho g}=4\cos\theta_c\frac{\sigma}{D^2\rho g} \tag{2.237}$$

where θ_c is the contact angle (°).

As mentioned above, both the Lord Rayleigh method and the Buckingham Π theorem generate dimensionless parameters automatically. However, the number of dimensionless parameters depends solely on the initial choice of the dimensional quantities responsible for the phenomenon. Their adequacy is therefore not guaranteed. Deep insight into the flow phenomenon is a key determinant for adequate dimensionless parameters.

2.7.3 Physical Meaning of Dimensionless Numbers

2.7.3.1 Reynolds Number

The Reynolds number, *Re*, is defined as

$$Re=\frac{VL}{\nu_f} \tag{2.238}$$

where V is the representative or characteristic velocity (m/s), L is the representative or characteristic length of the flow field (m), and ν_f is the kinematic viscosity of fluid (m^2/s). Equation (2.238) can be rewritten as

$$Re=\frac{VL}{\nu_f}\rightarrow\frac{(\rho V^2/2)L^2}{\rho\nu_f\frac{V}{L}L^2}=\frac{(\rho V^2/2)L^2}{\mu\frac{V}{L}L^2}=\frac{(\rho V^2/2)L^2}{\tau L^2}\left(\frac{\text{Inertial force}}{\text{Viscous force}}\right) \tag{2.239}$$

where ρ is the density of the fluid (kg/m^3), μ is the dynamic viscosity of the fluid (kg/m s), and τ is the shear stress (Pa). Thus, the Reynolds number represents the ratio of the inertial force to the viscous force in the flow field. As *Re* increases, the inertial force becomes dominant and, hence, transition to turbulence occurs.

2.7.3.2 Mach Number

The Mach number is expressed by

$$M = \frac{V}{c} \tag{2.240}$$

where c is the speed of sound (m/s). This dimensionless number can be rewritten as

$$M = \left(\frac{\rho V^2/2}{\rho c^2/2}\right)^{1/2} \tag{2.241}$$

where $\rho V^2/2$ and $\rho c^2/2$ denote the kinetic and elastic energies, respectively. The ratio of the kinetic energy to the elastic energy of fluid is called the Cauchy number. Accordingly, the Mach number is the root-mean-square value of the Cauchy number.

2.7.3.3 Strouhal Number

The Strouhal number is defined as

$$St = \frac{fL}{V} \tag{2.242}$$

where f $(=1/T)$ is the representative frequency and T is the representative time.

According to the Navier–Stokes equation in the x direction Eq. (2.25), described in a previous Sect. 2.1.2, the component terms are expressed by

$$\text{Temporal acceleration}: \ \partial u/\partial t \tag{2.243}$$

$$\text{Spatial acceleration}: \ u\partial u/\partial x + v\partial u/\partial y + w\partial u/\partial z \tag{2.244}$$

where the velocity components in the x, y, and z directions are denoted by u, v, and w, respectively, and t is time.

We consider the ratio of $\partial u/\partial t$ to $u\partial u/\partial x$ and introduce a representative velocity, V, and a representative length, L, in addition to the representative time, T, described above in order to non-dimensionalize u, x, and t thus:

$$u' = \frac{u}{V} \tag{2.245}$$

$$x' = \frac{x}{L} \tag{2.246}$$

$$t' = \frac{t}{T} \tag{2.247}$$

The ratio is expressed as

$$\frac{\partial u/\partial t}{u\partial u/\partial x} = \frac{(V/T)\partial u'/\partial t'}{(V^2/L)u\partial u'/\partial x'} = \frac{L}{VT}\frac{\partial u'/\partial t'}{u'\partial u'/\partial x'} \tag{2.248}$$

As the magnitude of $\partial u'/\partial t'/(u'\partial u'/\partial x')$ is on the order of unity, O(1), the value of $\partial u/\partial t/(u\partial u/\partial x)$ is governed by $L/(VT)$. Here, $L/(VT)$ is rewritten as

$$\frac{L}{VT} = \frac{fL}{V} \equiv St\left(\frac{\text{Temporal accelaration}}{\text{Spatial accelaration}}\right) \tag{2.249}$$

Consequently, the Strouhal number represents the ratio of the temporal acceleration to the spatial acceleration.

2.7.3.4 Weber Number

$$We = \frac{\rho L V^2}{\sigma} \rightarrow \frac{(\rho V^2/2)L^2}{\sigma L}\quad\left(\frac{\text{Inertial force}}{\text{Surface tension force}}\right) \tag{2.250}$$

where ρ is the density of fluid (kg/m^3), L is the representative length (m), V is the representative velocity (m/s), and σ is the surface tension of fluid (N/m).

2.7.3.5 Froude Number

This dimensionless parameter was originally proposed by Froude in ship engineering.

$$Fr = \frac{V}{(gL)^{1/2}} \rightarrow \left[\frac{(\rho V^2/2)L^2}{\rho g L^3}\right]^{1/2}\quad\left(\frac{\text{Inertial force}}{\text{Buoyancy force}}\right)^{1/2} \tag{2.251}$$

where g is the acceleration due to gravity.

2.7.3.6 Knudsen Number

$$Kn = \frac{\lambda_{\mathrm{m}}}{L} \tag{2.252}$$

where λ_{m} is the mean free path.

References

1. Schlichting H (1979) Boundary-layer theory (translated by J. Kestin, 2nd edn.) McGraw-Hill New York
2. Bird RB, Stewart WE, Lightfoot EN (1960) Transport phenomena. Wiley, New York
3. Poirier DR, Geiger GH (1994) Transport phenomena in materials processing. TMS, Warrendale
4. Batchelor GK (2000) An introduction to fluid dynamics. Cambridge University Press, Cambridge
5. Streeter VL, Wylie EB (1985) Fluid mechanics, 8th edn. McGraw-Hill, New York
6. Itaya M (1978) Fluid mechanics. Asakura Book, Co., Ltd., Tokyo

Chapter 3
Special Topics in Fluid Mechanics

Abstract Three main targets of this chapter are non-Newtonian fluid flows, microscale and nanoscale fluid mechanics, and mixing of fluids. Discussion is given mainly from a practical point of view.

Keywords Apparent dynamic viscosity • Fuel cell • Micro channel • Micro reactor • Mixing • Non-Newtonian fluid • Rheology

3.1 Rheology

3.1.1 Definition of Rheology

Rheology, originally attributed to Bingham, considers the deformation and flow of substances [1]. In other words, rheology focuses on the constitutive relationship between the stress and strain of a substance. Elastic theory on the other hand considers the deformation of a substance on the basis of Hook's law while fluid mechanics is mainly concerned with fluid flow on the basis of Newton's law of viscosity. Fluids that obey Newton's law are called Newtonian fluids. Newtonian fluids are quite common and include air, water, oil, and molten metal. Other fluids that do not obey Newton's law are collectively called non-Newtonian fluids. Rheology also covers the motion of non-Newtonian fluids commonly seen in, for example, colloid chemistry and high-molecular chemistry. Blood is a type of non-Newtonian fluid referred to as shear thinning. Specifically, the viscosity decreases with the shear rate. Rheology which is concerned with biological fluids such as blood flow in arteries and deformation of blood cells is classified as biorheology [2].

M. Iguchi and O.J. Ilegbusi, *Basic Transport Phenomena in Materials Engineering*,
DOI 10.1007/978-4-431-54020-5_3,

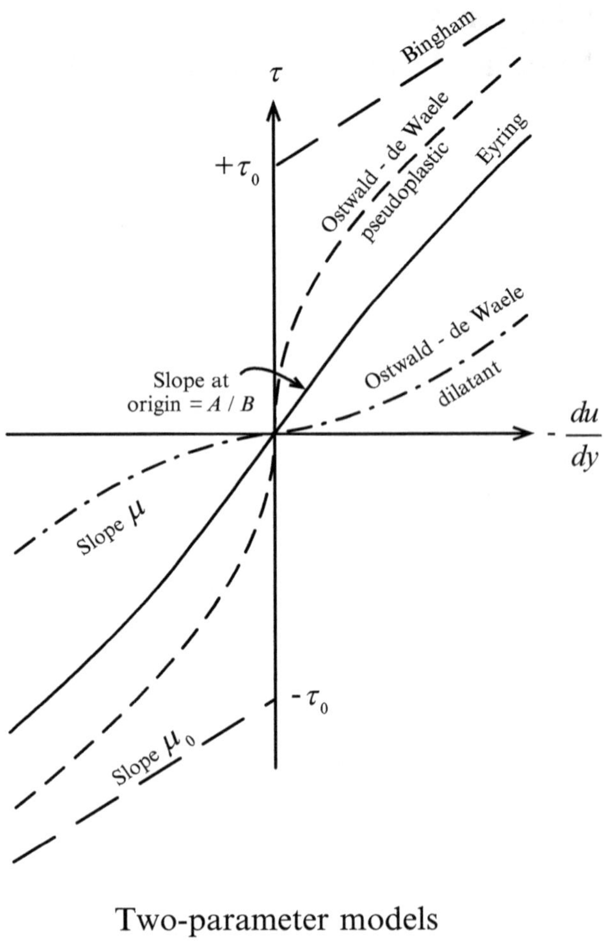

Fig. 3.1 Relationship between shear stress and velocity gradient for Newtonian and non-Newtonian fluids

3.1.2 Classification of Non-Newtonian Fluids

The shear stress in Newtonian flow follows the following simple relationship (see Fig. 3.1):

$$\tau = -\mu \frac{du}{dy} \tag{3.1}$$

where μ is the dynamic viscosity of fluid which is independent of the rate of shear, i.e., the velocity gradient, *du*/*dy*. The minus sign on the right-hand side of Eq. (3.1) is inserted when the velocity gradient is negative.

In non-Newtonian fluid flows Eq. (3.1) does not hold and other types of relationships have been proposed between shear stress and velocity gradient. Some of these relations are described below.

3.1.2.1 Bingham Model

$$\tau = -\mu_0 \frac{du}{dy} \pm \tau_0 \quad |\tau| > \tau_0 \tag{3.2}$$

$$\frac{du}{dy} = 0 \qquad |\tau| < \tau_0 \tag{3.3}$$

The positive sign on the right-hand side of Eq. (3.2) is used when τ is positive, and the minus sign is used when τ is negative. This model is known to be applicable to suspensions and pastes.

3.1.2.2 Ostwald–de Waele Model (Power Law)

$$\tau = -m\left|\frac{du}{dy}\right|^{n-1} \frac{du}{dy} \tag{3.4}$$

The values of coefficients m and n depend on the type of fluid and are fully discussed in Bird et al. [1]. For $m = \mu$ and $n = 1$ Eq. (3.4) reduces to Newton's law of viscosity.

3.1.2.3 Eyring Model

$$\tau = A \sinh^{-1}\left(-\frac{1}{B}\frac{du}{dy}\right) \tag{3.5}$$

where A and B are constants. Further information on this model can be found in Bird et al. [1].

3.1.3 Apparent Dynamic Viscosity, μ_{app}

In practical applications the apparent dynamic viscosity is commonly used for describing the viscosity of non-Newtonian fluids such as suspensions [3–4]. The apparent (or effective) dynamic viscosity, μ_{app} (kg/(m s)), is defined from the relationship

$$\tau = -\mu_{\text{app}} \frac{du}{dy} \tag{3.6}$$

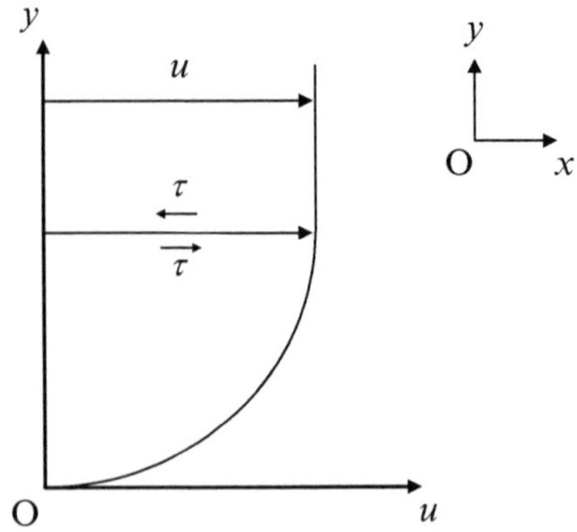

Fig. 3.2 Shear stress in suspension flow

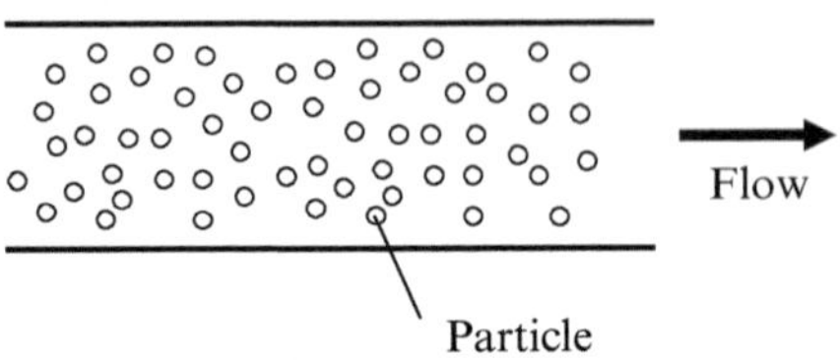

Fig. 3.3 Pipe flow containing many particles

where τ (Pa) is the shear stress, u (m/s) is the fluid velocity in the flow (x) direction, and y (m) is the coordinate perpendicular to the x direction, as can be seen in Fig. 3.2. It should be noted that the apparent viscosity was originally introduced by assuming that the flow containing suspensions behaves as a single-phase Newtonian fluid flow.

In materials engineering the apparent dynamic viscosity, μ_{app}, is often used in the casting process to describe the behavior of semisolid metal slurries. In addition, it is applicable to liquids containing small particles, i.e., suspensions. A variety of theoretical and empirical equations have been proposed for suspensions of solid spheres (see Fig. 3.3).

3.1.3.1 Einstein (Very Dilute State for Spheres)

$$\frac{\mu_{\text{app}}}{\mu_0} = 1 + 2.5\varphi \tag{3.7}$$

where μ_0 (kg/m s) is the dynamic viscosity of fluid without spheres and φ (−) is the volume fraction of solid spheres.

3.1.3.2 Taylor Model

$$\frac{\mu_{\text{app}}}{\mu_0} = 1 + \frac{2.5\varphi + 1}{\varphi + 1} \tag{3.8}$$

3.1.3.3 Einstein–Roscoe Model

$$\frac{\mu_{\text{app}}}{\mu_0} = \left(1 - \frac{\varphi}{\varphi_C}\right)^{-n} \tag{3.9}$$

This relationship is typically used when spheres of equal size are dispersed, $\varphi_C = 0.74$ and $n = 2.5$. Other empirical relations are also used for suspensions with liquid droplets including the following.

3.1.3.4 Vermeulen et al. [4]

$$\frac{\mu_{\text{app}}}{\mu_c} = \frac{1}{1-\varphi}\left(1 + \frac{1.5\varphi\mu_d}{\mu_c + \mu_d}\right) \tag{3.10}$$

where μ_d is the dynamic viscosity of dispersed phase (kg/m s), φ is the volume fraction of droplets (−), and μ_c is the dynamic viscosity of continuous phase (kg/m s).

3.1.3.5 Water–Organic Liquid System

$$\frac{\mu_{\text{app}}}{\mu_w} = \frac{1}{1-\varphi_w}\left(1 + \frac{6\varphi_o\mu_o}{\mu_w + \mu_o}\right) \quad (\varphi_w > 0.4) \tag{3.11}$$

where μ_w is the dynamic viscosity of water (kg/m s), φ_w is the volume fraction of water (−), μ_o is the dynamic viscosity of oil (kg/m s), and φ_o is the volume fraction of oil (−).

3.2 Flow in Porous Media

3.2.1 Packed Bed or Packed Column

A flow passage containing various types of particles is regarded as a packed bed. Such a system bed is prevalent in chemical and materials engineering (see Fig. 3.4).

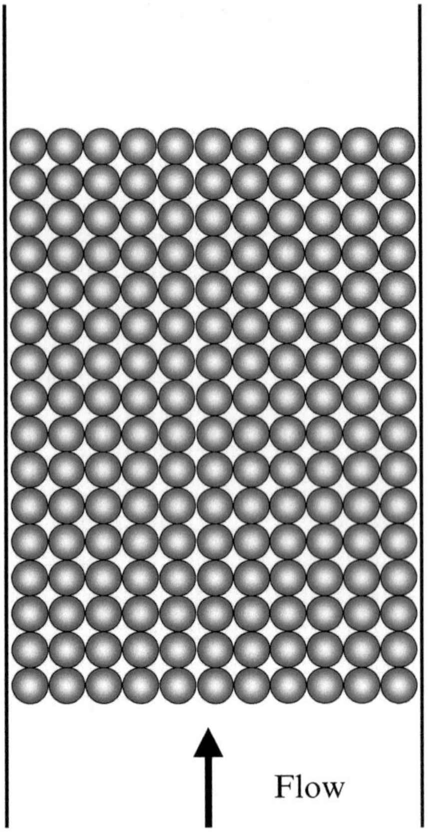

Fig. 3.4 Flow in packed bed

3.2.2 *Pressure Drop in a Packed Bed with Spherical Particles*

3.2.2.1 Blake–Kozeny Equation for Laminar Flow [1]

$$\Delta p = \lambda \frac{L}{d_\mathrm{p}} \frac{\rho {v_\mathrm{m}}^2}{2} \tag{3.12}$$

$$\lambda = \frac{300(1-\varepsilon)^2}{\varepsilon^3} \frac{\nu_\mathrm{f}}{d_\mathrm{p} v_\mathrm{m}} \quad \left(\frac{(1-\varepsilon) d_\mathrm{p} v_\mathrm{m}}{\nu_\mathrm{f}} < 10 \right) \tag{3.13}$$

where L is the height of bed, d_p is the diameter, ρ is the fluid density, v_m is the cross-sectional mean velocity, ν_f is the kinematic viscosity, and ε is the porosity.

3.2.2.2 Blake–Kozeny Equation for Turbulent Flow

$$\lambda = 3.5\frac{1-\varepsilon}{\varepsilon^3}\left(\frac{d_p v_m}{(1-\varepsilon)\nu_f} > 1{,}000\right) \tag{3.14}$$

3.2.3 *Ergun Equation*

Combining Eqs. (3.12) and (3.14) Ergun derived the following equation applicable to both laminar and turbulent flows:

$$\frac{\Delta p}{\rho {v_m}^2}\left(\frac{d_p}{L}\right)\left(\frac{\varepsilon^3}{1-\varepsilon}\right) = 150\frac{(1-\varepsilon)\nu_f}{d_p v_m} + 1.75 \tag{3.15}$$

3.2.4 *Pressure Drop in Packed Bed with Nonspherical Particles*

The following specific area of a particle, A_{sp}, is introduced:

$$A_{sp} = \frac{A_p}{V_p} \tag{3.16}$$

where A_p and V_p are the surface area and volume of the particle, respectively (see Fig. 3.5). For example, A_{sp} is expressed for a sphere as

$$A_{sp} = \frac{A_p}{V_p} = \frac{\pi {D_p}^2}{\pi {D_p}^3/6} = \frac{6}{D_p} \tag{3.17}$$

Here we further introduce the following specific particle diameter for nonspherical particle:

$$D_{sp} = \frac{6}{A_{sp}} = \frac{6V_p}{A_p} \tag{3.18}$$

This diameter should be substituted into Eqs. (3.12), (3.14), and (3.15) when the pressure drop prediction is necessary for a packed bed containing nonspherical particles.

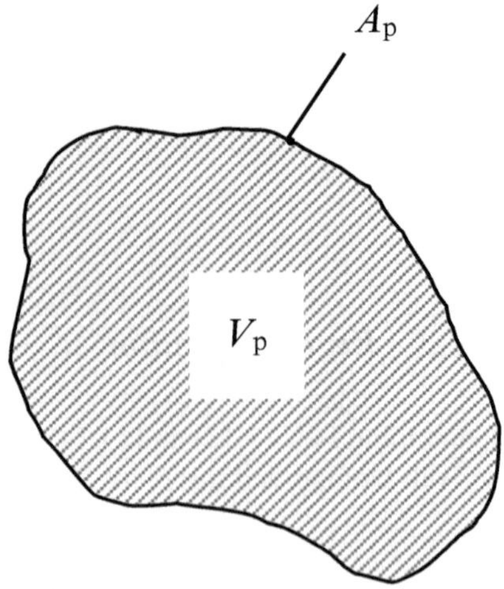

Fig. 3.5 Surface area, A_p, and volume, V_p, of a nonspherical body

Exercise 3.1 Obtain the pressure loss, Δp, of airflow in a packed bed of 20 cm in diameter and 40 cm in height using the Ergun equation. The bed is packed with solid spheres of diameter, $D_p = 1.0$ mm. The clearance velocity of air, v_{mc}, is 1.0 m/s, the porosity, ε, is 0.40, and the kinematic viscosity, ν_f, is 1.5×10^{-5} m^2/s.

Answer to Exercise 3.1 The pressure drop can be calculated from Eq. (3.15) as follows:

$$\begin{aligned}\Delta p &= \left(150\frac{(1-\varepsilon)\nu_f}{d_p v_m} + 1.75\right)\left(\frac{L}{d_p}\right)\left(\frac{1-\varepsilon}{\varepsilon^3}\right)\rho {v_m}^2 \\ &= \left(150\frac{(1-0.4)\times 1.5\times 10^{-5}}{1.0\times 10^{-3}\times 1.0} + 1.75\right)\left(\frac{0.4}{1.0\times 10^{-3}}\right)\left(\frac{1-0.4}{(0.4)^3}\right)\times 1.23\times (1.0)^2 \\ &= 14.3\,\text{kPa}\end{aligned} \tag{3.19}$$

3.3 Microscale and Nanoscale Fluid Mechanics

With the rapid development of chemical and biomedical engineering, information on flows in microscale and nanoscale pipelines and reactors are increasingly important. In this chapter particular focus is on micro reactors, in order to introduce the relevant concepts in small scale systems. Additional information is provided in a subsequent Part IV on practical applications of micro channels.

3.3.1 Flow in Micro Reactor

Micro reactors are frequently used for effectively producing desirable materials and analyzing substances with small volume [5–8]. As mentioned in the previous Sect. 3.2.2, a fluid can be regarded as a continuum when the following condition is satisfied:

$$Kn = \frac{\lambda_{\mathrm{m}}}{L} < \frac{1}{5} \tag{3.20}$$

where Kn is the Knudsen number, λ_{m} is the mean free path, and L is the representative length. The mean free path, λ_{m}, for a gas is calculated from the theory to give

$$\lambda_{\mathrm{m}} = \frac{1}{\sqrt{2}\pi {d_{\mathrm{m}}}^2 (N_{\mathrm{m}}/V)} \tag{3.21}$$

where d_{m} is the diameter of a molecule, N_{m} is the number of molecules, and V is the volume of the space considered. Accordingly, N_{m}/V denotes the number of molecules per unit volume [6].

For example, under atmospheric condition the mean free path, λ_{m}, for air is given by

$$\begin{aligned} \lambda_{\mathrm{m}} &= \frac{1}{\sqrt{2}\pi {d_{\mathrm{m}}}^2 (N_{\mathrm{m}}/V)} = \frac{1}{\sqrt{2}\pi \times (0.4 \times 10^{-9})^2 \times 6.03 \times 10^{23}/0.0224} \\ &= 5.2 \times 10^{-8}\,\mathrm{m} \end{aligned} \tag{3.22}$$

where $N_{\mathrm{m}} = 6.03 \times 10^{23}$ is the Avogadro number.

Substituting Eq. (3.22) into Eq. (3.20) yields

$$L > 5\lambda_{\mathrm{m}} = 5 \times 5.2 \times 10^{-8}\ \mathrm{m} = 2.6 \times 10^{-7}\,\mathrm{m} = 0.26\,\mu\mathrm{m} \tag{3.23}$$

Thus, if the representative length, L, is greater than 0.25 μm, air can be regarded as a continuum and the conventional fluid mechanics principle is applicable to micro reactors. Of course, the mean free path for liquids is much shorter than that for gas.

3.3.2 Fluid Flow Characteristics in Micro Reactor

3.3.2.1 Flow Regime

As the representative length, L, becomes very small in micro reactors, flows are usually laminar in nature, as will be confirmed below. The Reynolds number is

$$Re = \frac{VL}{\nu_{\mathrm{f}}} \tag{3.24}$$

where V is the representative velocity and ν_f is the kinematic viscosity of the fluid. The maximum values of the representative velocity and length are assumed to be approximately $V = 1.0\times10^{-3}$ m/s and $L = 1 \times 10^{-4}$ m, respectively [5]. If the fluid is water, its kinematic viscosity is approximately 1.0×10^{-6} m^2/s. Equation (3.24) gives the following Reynolds number:

$$Re = \frac{VL}{\nu_f} = \frac{1.0 \times 10^{-3} \times 1.0 \times 10^{-4}}{1 \times 10^{-6}} = 0.10 \tag{3.25}$$

This flow is laminar because the Reynolds number thus calculated is much smaller than the critical value of 2,320.

3.3.2.2 Surface Area to Volume Ratio (Surface-to-Volume Ratio)

The inner surface area of a circular pipe, A_{pipe}, and its volume, V_{pipe}, are respectively expressed by

$$A_{\text{pipe}} = \pi DL \tag{3.26}$$

$$V_{\text{pipe}} = \pi D^2 L/4 \tag{3.27}$$

where D is the pipe diameter and L is the length of the pipe. The ratio of the surface area, A_{pipe}, to the volume of the pipe, V_{pipe}—usually called the surface-to-volume ratio—is given by

$$R_{\text{av}} = \frac{\pi DL}{\pi D^2 L/4} = \frac{4}{D} \tag{3.28}$$

This equation implies that the surface-to-volume ratio, R_{av}, increases in inverse proportion to the pipe diameter, D. This trend is beneficial for the enhancement of chemical reactions.

3.4 Surface Forces

3.4.1 *Role of Surface Tension Force*

The effects of the so-called body forces such as the buoyancy force, gravitational force, and centrifugal force become relatively weak with decrease in the size of a micro reactor compared to the surface tension force and viscous force. In other words, the Reynolds number and the Weber number become small.

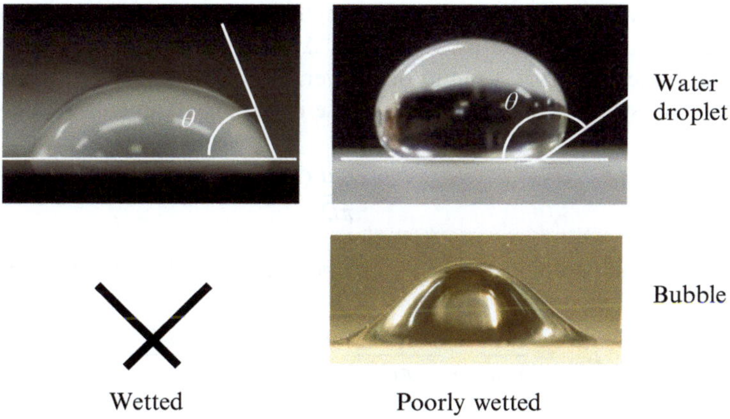

Fig. 3.6 Definition of contact angle

In addition, wettability plays an important role in micro reactors. The wettability is quantitatively described by the equilibrium contact angle, θ_c, as already mentioned. Figure 3.6 shows photographs of a sessile droplet placed on a horizontal solid plate. When the contact angle, θ_c, is less than 90° ($0 \leq \theta_c < 90°$), the plate is wetted by the droplet, while it is poorly wetted by the droplet for $90 \leq \theta_c \leq 180°$. If a poorly wetted plate is immersed in a liquid pool, a bubble can be trapped on the upper side of the plate. The height of the bubble becomes lower and the bubble spreads more widely as the contact angle increases. When the pipeline system of a micro reactor is poorly wetted by a liquid, the inner walls of the system are preferably covered with gas instead of the liquid, and blockage of the pipe often occurs due to attachment of bubbles. Fujii et al. [5] used this phenomenon effectively to divide a liquid column into droplets of predetermined volume continuously. A droplet can be moved on a flat plate without any external force by applying a wettability gradient on the plate. This result could be useful for driving liquid in micro reactors in the near future. Other surface forces must be considered depending on the scale of the micro reactor [5–7]. These include van der Waals forces, electrostatic forces, and steric forces.

3.4.2 Flow Control in Micro Reactors

3.4.2.1 Pump and Blower

The pressure loss in laminar flow in a circular pipe is given by

$$\Delta p = \frac{128\mu LQ}{\pi D^4} \tag{3.29}$$

where μ is the dynamic viscosity, L is the length of the pipe, and Q is the flow rate. This equation states that the pressure loss, Δp, is inversely proportional to D^4. Accordingly, the power of a pump or blower, $P_w = (\Delta p Q)$, called the pumping power, increases significantly with a decrease in the pipe diameter, D, thus:

$$P_w = \frac{128\mu L Q^2}{\pi D^4} \tag{3.30}$$

The ratio of the pumping power, P_w, to the volume of the pipe, V_{pipe}, is

$$R_{pv} = \frac{P_w}{V_{pipe}} = \frac{128\mu L Q^2}{\pi D^4} \times \frac{4}{\pi D^2} = \frac{512\mu L Q^2}{\pi^2 D^6} \tag{3.31}$$

The pumping power per unit pipe volume therefore is inversely proportional to D^6. Equation (3.31) is rewritten as a function of the cross-sectional mean velocity, v_m, thus:

$$R_{pv} = \frac{512\mu L Q^2}{\pi^2 D^6} = \frac{32\mu L {v_m}^2}{D^2} \tag{3.32}$$

It should be noted that although the surface-to-volume ratio, R_{av}, actually increases with a decrease in the pipe diameter, D, the pumping power, P_w, increases much more significantly. Additional details on micro pump are available elsewhere [5, 6].

3.4.2.2 Electrophoresis Method

As mentioned above, driving a fluid in micro reactors needs considerable amount of power, P_w. The electrophoresis method is considered an alternative approach. When a certain voltage is supplied to liquid placed adjacent to a solid wall, flow is induced along the wall, called electrophoresis flow. This phenomenon is quite useful for driving liquid in micro reactors.

3.4.2.3 Capillary Force

Capillary force could be effectively used in micro channels. For example, surface tension gradient in the flow direction could control the motion of a bubble in micro channel.

3.4.2.4 Mixing in Micro Reactor

A variety of mixing methods are applied in the conventional-scale reactors as described below:

Mechanical Agitation Using Impeller

Mechanical agitation is extensively used in many engineering fields such as chemical and mechanical engineering. This method is seldom used in materials engineering because of potential damage to the impeller due to erosion as soon it is immersed in a high-temperature molten metal bath. An exception is the KR process used in dephosphorization [9–11].

Gas Injection

Gas is typically injected into molten metal bath in metals refining processes from a bottom nozzle, a top lance, or a side nozzle. In all cases the gas-lift effect is used for agitating the bath.

Liquid Injection

A liquid jet is introduced into a bath of the same liquid or of a different liquid. The liquid surrounding the jet is entrained into the jet. This entrainment causes a recirculation motion which agitates the liquid in the bath.

Electromagnetic Agitation

The Lorentz (electromagnetic) force is often used for agitating molten metal in materials processing operations. No agitating device is immersed in a molten metal bath and, hence, contamination of the metal can be prevented.

Ultrasonic Wave Agitation

An ultrasonic wave is periodically applied to a liquid to generate cavitation bubbles. Periodical generation and disappearance of the bubbles are closely associated with bath mixing.

Plunging Jet Agitation

A water fall is a type of plunging jet. Another example is hot water jet issuing out of a jar which is used for mixing coffee powder and cream contained in a cup. The plunging jet is considered one of the promising candidates for effectively agitating molten metal.

Static Mixer

A static mixer is widely used in chemical engineering. This method allows some liquids to be mixed without using movable devices such as an impeller.

The above methods cannot be effectively used in micro reactors. Mixing of fluids in such cases is accomplished mainly through diffusion.

3.4.2.5 Measurement of Fluids Flow in Micro Reactor and Micro Channel

Recently, many types of micro-PIV (particle image velocimetry) or techniques have been developed to measure the velocity fields in micro reactors and micro channels [12–18]. The measurement principle of PIV is described in other texts [18–20].

3.4.2.6 Numerical Simulation

When a flow is laminar and regarded as a continuum in a micro channel, predicting the characteristics of the flow is possible on the basis of the conventionally used numerical simulation methods. This is true even if the channels are very complicated in shape and size [21].

3.4.3 Applications

3.4.3.1 Microelectromechanical Systems (MEMS)

In fluids engineering MEMS is widely applied to micro pumps, micro valves, cell sorters, biosensors, microanalysis tips, and the like [22–32].

3.4.3.2 Sheath Flow

A typical sheath flow is embodied in flow cytometry which can count the number of red blood cells, white blood cells, and so on [7]. A micro channel is covered with a sheath, and the blood cells flow in the micro channel. The cells are illuminated with a laser beam after exiting the micro channel. The number of cells can be counted by analyzing the beam reflected from the cells [31, 32].

References

1. Bird RB, Stewart WE, Lightfoot EN (1960) Transport phenomena. Wiley, New York
2. Oka S (1976) Rheology biorheology, 2nd edn. Shokabo, Tokyo
3. Hirai M, Takebayashi K, Yoshikawa Y, Yamaguchi R (1992) Apparent viscosity of semi-solid metals. Tetsu-to-Hagane 78(6):902–909
4. Michaelides EE (2006) Particles, bubbles and drops. World Scientific, New Jersey, p 304
5. Fujii T (2008) Development and applications of micro reactor. In: Yoshida Y, CMC, (eds) Tokyo, pp 47–60
6. Kusakabe K, Sotowa K (2008) Introduction to micro reactor. Yoneda Pub. Co. Ltd., Tokyo, pp 43–78
7. Togashi S, Miyake R (2005) Micro reactor technology. NTS 41–49
8. Kawai T (2003) Dictionary of nanotechnology. Kogyo-chosakai, Tokyo, 527
9. Sahai Y, Pierre GR (1992) Advances in transport processes in metallurgical systems. Elsevier, Amsterdam
10. Ilegbusi OJ, Iguchi M, Whanshiedler W (1999) Mathematical and physical modeling of materials processing operations. Chapman-Hall/CRC, Boca Raton
11. Mazumdar D, Evans JW (2010) Modeling of steelmaking processes. CRC, Boca Raton
12. Schroeder A, Willert CE (2008) Particle image velocimetry. Springer, Berlin
13. Kimura I, Uemura T, Okuno T (2001) Visualized information analysis. Kindaikagaku Co., Ltd., Tokyo
14. Nyguen CV, Fouras A, Carberry J (2010) Improvement of measurement accuracy in micro PIV by image overlapping. Exp Fluids 49:701–712
15. Sugii Y, Okamoto K (2004) Quantitative visualization of micro-tube flow using micro-PIV. J Vis 7–1:9–16
16. Suzuki T, Teramura Y, Hata H, Inokuma K, Kanno I, Iwata H, Kotera H (2007) Development of a micro biochip integrated traveling wave micropumps and surface Plasmon resonance imaging sensors. Microsyst Technol 13:1391–1396
17. Sheen HJ, Hsu CJ, Wu TH, Chang CC, Chu HC, Yang CY, Lei U (2008) Velocity measurements of blood flow in a rectangular PDMS microchannel assessed by confocal micro-PIV system. Microfluid Nanofluid 4:331–342
18. Wereley AT, Meinhart CD (2010) Recent advances in micro-particle image velocimetry. Ann Rev Fluid Mech 42:557–576
19. Yang WJ (2001) Handbook of flow visualization, 2nd edn. Taylor and Francis, New York
20. Visualization Society of Japan (2002) Handbook of particle image velocimetry. Morikita, Tokyo
21. Japan Society of Mechanical Engineers (1986) Handbook of mechanical engineering, A5, fluids engineering. JSME, Tokyo
22. Ho CM, Tai YC (1998) Micro-electro-mechanical-systems (MEMS) and fluid flows. Ann Rev Fluid Mech 30:579–612
23. Beebe DJ, Mensing GA, Walker GM (2002) Physics and applications of microfluidics in biology. Ann Rev Biomed Eng 4:261–286
24. Kasagi N, Suzuki Y, Fukagata K (2009) Microelectromechanical systems-based feedback control of turbulence for skin friction reduction. Ann Rev Fluid Mech 41:231–251
25. Gutierrez-Osuna R, Hierlemann A (2010) Adaptive microsensor systems. Ann Rev Anal chem 3:255–276
26. Stone HA, Stroock AD, Ajdari A (2004) Engineering flows in small devices: microfluidics toward a lab-on-a-chip. Ann Rev Fluid Mech 36:381–411
27. Rauscher M, Dietrich A (2008) Wetting phenomena in nanofluidics. Ann Rev Mater Res 38:143–172
28. Toner M, Irimia D (2005) Blood-on-a-chip. Ann Rev Biomed Eng 7:77–103
29. McCalla SE, Tripathi A (2011) Microfluidic reactors for diagnostics applications. Ann Rev Biomed Eng 13:321–343

30. Ghosal S (2006) Electrokinetic flow and dispersion in capillary electrophoresis. Ann Rev Fluid Mech 38:309–338
31. Livak-Dahl E, Sinn I, Burns M (2011) Microfluidic chemical analysis system. Ann Rev Chem Bio Eng 2:325–353
32. Meyvantsson I, Beebe DJ (2008) Cell culture models in microfluidic systems. Ann Rev Anal Chem 1:423–449

Part II
Single-Phase Heat Transfer

Chapter 4
Basic Principles of Heat Transfer

Abstract This chapter deals with three different types of heat transfer: conduction, convection, and radiation heat transfer. Relevant dimensionless numbers such as Nusselt number, Prandtl number, and Grashof number are shown to help systematically understand the heat transfer.

Keywords Conduction • Convection • Fourier's law • Newton's law of cooling • Nusselt number • Prandtl number • Radiation

4.1 Introduction

Materials processes typically occur at high temperature, often in excess of 1,000 °C. Thus, such processes involve heat transfer which may be accompanied by significant heat loss. A thorough understanding of heat transfer is therefore essential in materials processing operations.

According to Holman [1], heat transfer is that science which seeks to predict the energy transfer that occurs between material bodies as a result of temperature difference. Heat transfer can be classified into three broad categories, as schematically shown in Fig. 4.1 [1–10].

4.1.1 Conduction

When a temperature gradient exists in a body, there is an energy transfer from the high-temperature region to the low-temperature region. This type of heat transfer is called conduction heat transfer.

M. Iguchi and O.J. Ilegbusi, *Basic Transport Phenomena in Materials Engineering*,
DOI 10.1007/978-4-431-54020-5_4,

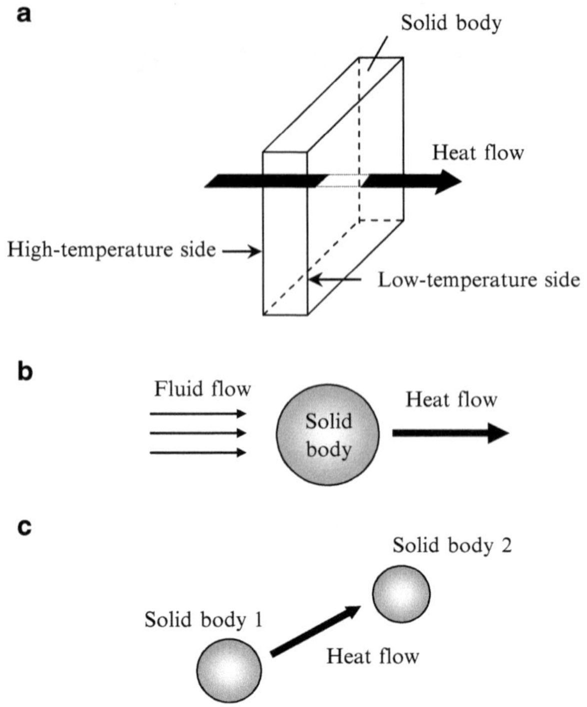

Fig. 4.1 Three types of heat transfer. (**a**) Conduction, (**b**) convection, (**c**) radiation

4.1.2 Convection

Heat transfer is accompanied by a fluid motion.

4.1.3 Radiation

Radiation is an electromagnetic radiation emitted by a body as a result of its temperature.

In Part II of this book heat transfer due to phase change also will be introduced in addition to the above three modes, using examples.

4.2 Conduction

Conduction heat transfer is described by Fourier's law (see Fig. 4.2):

$$Q = -\lambda \frac{d\theta}{dx} At \tag{4.1}$$

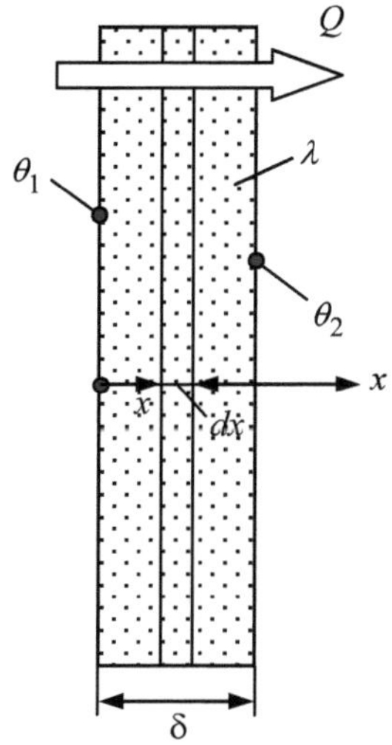

Fig. 4.2 One-dimensional heat conduction

where Q is the heat transferred by conduction (J), λ is the thermal conductivity (W/m K), θ is the temperature (K), x is the distance in the direction of the heat flow (m), A is the area of the body projected to a plane normal to the direction of the heat flow (m^2), and t is time (s). The minus sign is inserted on the right-hand side of Eq. (4.1) so that the second principle of thermodynamics will be satisfied. This allows heat to be transferred down the temperature gradient, $d\theta/dx$.

Dividing both sides of Eq. (4.1) by At gives

$$q = -\lambda \frac{d\theta}{dx} \tag{4.2}$$

where q is the heat flux, i.e., heat flow per unit time and unit area. This equation indicates that the heat flux, q, is proportional to the temperature gradient, $d\theta/dx$.

The reader may notice that a similar relationship exists in fluid flow as Newton's law of viscosity,

$$\tau = -\mu \frac{dv}{dx} \tag{4.3}$$

where τ is the shear stress, μ is the dynamic viscosity, and v is the velocity of fluid. This relationship is known as Newton's law of viscosity. The unit of shear stress is (Pa or N/m^2) and it can be transformed as follows:

$$\left(N/m^2\right) = \left(kgm/s^2\right)/\left(m^2\right) = (kgm/s)/\left(s\,m^2\right) \tag{4.4}$$

In this equation (kgm/s) is the unit of momentum and, hence, (kgm/s)/(s m^2) is the momentum flux. Equation (4.3) therefore states that the momentum flux is proportional to the velocity gradient, dv/dx.

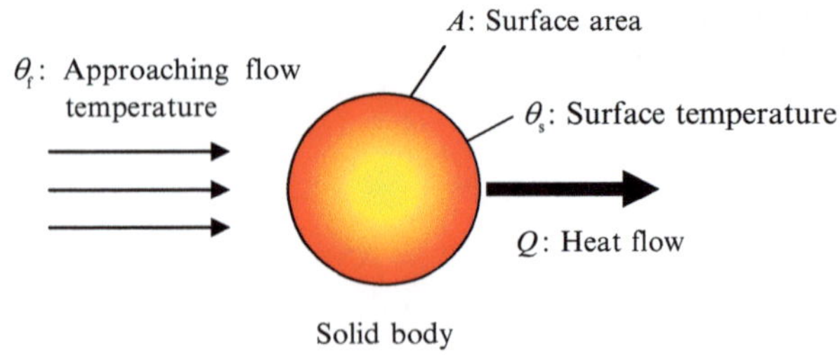

Fig. 4.3 Convection heat transfer

4.3 Convection

Newton's law of cooling is expressed as

$$Q = \alpha(\theta_s - \theta_f)At \tag{4.5}$$

where Q is the heat transferred by convection, α is the convection heat-transfer coefficient (abbreviated as the heat-transfer coefficient) (W/m^2 K), θ_s is the temperature of a body (K), θ_f is the temperature of a fluid approaching the body (K), A is the surface area of the body (m^2), and t is time (s) (see Fig. 4.3). It should be noted that the heat-transfer coefficient is sometimes denoted by h in place of α.

Dividing both sides of Eq. (4.5) by At gives

$$q = \alpha(\theta_s - \theta_f) \tag{4.6}$$

The heat flux, q, therefore is proportional to the temperature difference $(\theta_s - \theta_f)$.

4.4 Radiation

Radiation heat transfer is basically described by the Stefan–Boltzmann law. This mode of heat transfer will be described in detail in Chap. 7.

4.5 Hydrodynamic Boundary Layer and Thermal Boundary Layer

According to Holman [1], just as the hydrodynamic boundary layer is defined as that region of the flow where viscous forces are felt, a thermal boundary layer may be defined as that region where temperature gradients are present in the flow. These temperature gradients would result from a heat-exchange process between the fluid and the wall.

Description of the hydrodynamic boundary layer and thermal boundary layer will be given with reference to Stokes' first problem [11]. This problem deals with flow around a wide, flat plate starting suddenly from rest at a constant nonzero velocity V. The fluid adjacent to the plate begins to move with it with time. This layer is called a hydrodynamic boundary layer. The velocity of fluid decreases in the direction normal to the plate from V on the surface of the plate to zero outside the layer. Its thickness, δ_{h}, increases with time, t, in the following manner:

$$\delta_{\mathrm{h}} = 4(\nu_{\mathrm{f}} t)^{1/2} \tag{4.7}$$

where ν_{f} ($\mathrm{m^2/s}$) is the kinematic viscosity of the fluid. The growth of the hydrodynamic boundary layer therefore is closely associated with the kinematic viscosity, ν_{f}. In other words, the kinematic viscosity is a measure of the degree of momentum exchange in the fluid flow.

In a similar manner, the thickness of a thermal boundary layer on a flat plate is governed by the thermal diffusivity of fluid, κ_{f} ($\mathrm{m^2/s}$); thus,

$$\kappa_{\mathrm{f}} = \lambda_{\mathrm{f}}/(\rho_{\mathrm{f}} c_{\mathrm{p}}) \tag{4.8}$$

where λ_{f} is the thermal conductivity of the fluid, ρ_{f} is the density of the fluid, and c_{p} is the specific heat at constant pressure. The unit of thermal diffusivity is equal to that of kinematic viscosity.

4.6 Relevant Dimensionless Numbers

The ratio of the kinematic viscosity, ν_{f}, to the thermal diffusivity, κ_{f}, is termed the Prandtl number, Pr:

$$Pr = \nu_{\mathrm{f}}/\kappa_{\mathrm{f}} \tag{4.9}$$

The Prandtl number is a direct measure of the ratio of the thicknesses of the hydraulic and the thermal boundary layers.

This dimensionless number is approximately 7 for water at a temperature of 20 °C, approximately 0.7 for air, and on the order of 0.01 for molten metals. Concerning a solid body moving in water, it is worth noting that the hydrodynamic boundary layer grows faster than the thermal boundary layer.

The ratio of the kinematic viscosity, ν_{f}, to the diffusion coefficient, D, is called the Schmidt number, Sc:

$$Sc = \nu_{\mathrm{f}}/D \tag{4.10}$$

References

1. Shiba K (1973) Thermodynamics, vol 10. Iwanami Book Co., Ltd., Tokyo p 26
2. Holman JP (1972) Heat transfer, 3rd edn. McGraw-Hill Kogakusha, Tokyo
3. Kobayashi K, Iida Y (1995) Transport phenomena, Newth edn. Asakura Book Co., Ltd., Tokyo
4. Nishikawa K, Kitayama N (1988) Heat transfer. Ohmu-sha Book Co., Ltd., Tokyo
5. Japan Society of Mechanical Engineers (1980) Data book on heat transfer, 3rd edn. JSME, Tokyo
6. Isshiki N, Kitayama N (1988) Heat transfer, 2nd edn. Morikita-shuppan Co., Ltd., Tokyo
7. Stephan K, Abdelsalam M (1980) Int J Heat Mass Transfer 23(1):73
8. JSME: JSME Data Book (2009) Heat transfer, 5th edn. JSME, Tokyo, pp 131–140
9. Bird RB, Stewart WE, Lightfoot EN (1960) Transport phenomena. Wiley, New York
10. Poirier DR, Geiger GH (1994) Transport phenomena in materials processing. TMS, Warrendale
11. Schlichting H (1979) Boundary-layer theory (translated by Kestin J, 7th edn). McGraw-Hill, New York

Chapter 5
One-Dimensional Heat Conduction

Abstract This chapter offers basic principles on one-dimensional heat conduction through a flat plate and the wall of a long pipe. Some equations for the overall heat-transfer coefficient are derived using Fourier's law.

Keywords Fourier's law • Grashof number • Nusselt number • Overall heat-transfer coefficient • Prandtl number

5.1 Fourier's Law

Figure 4.2 in the previous chapter illustrates a one-dimensional conduction heat-transfer system. The system is assumed to be in a steady state. The temperatures on the left-hand-side and right-hand-side surfaces of the wall are θ_1 and θ_2, respectively ($\theta_1 > \theta_2$). The heat flux is expressed by

$$q = -\lambda(d\theta/dx) \tag{5.1}$$

Equation (5.1) is transformed into

$$d\theta = -(q/\lambda)dx \tag{5.2}$$

This equation is solved subject to the following boundary conditions:

$$\theta = \theta_1 \text{ at } x = 0 \tag{5.3}$$

$$\theta = \theta_2 \text{ at } x = \delta \tag{5.4}$$

where δ is the thickness of the wall. Equation (5.2) is integrated to give

$$\theta = -(q/\lambda)x + C \tag{5.5}$$

where C is the integration constant.

M. Iguchi and O.J. Ilegbusi, *Basic Transport Phenomena in Materials Engineering*, DOI 10.1007/978-4-431-54020-5_5, © Springer Japan 2014

The temperature at any location x inside the plate, θ, can be derived from Eqs. (5.3) and (5.5):

$$\theta = -(q/\lambda)x + \theta_1 \tag{5.6}$$

The heat flux, q, is obtained from Eqs. (5.4) and (5.6):

$$q = (\lambda/\delta)(\theta_1 - \theta_2) \tag{5.7}$$

5.2 Heat-Transfer Coefficient

Consider the one-dimensional system shown in Fig. 5.1. The temperatures of fluids surrounding the wall are θ_a and θ_b:

Newton's law of cooling gives the following heat fluxes:

(Left-hand side)

$$Q = \alpha_1(\theta_a - \theta_1)At \tag{5.8}$$

$$\frac{Q}{At} = q = \alpha_1(\theta_a - \theta_1) \tag{5.9}$$

(Right-hand side)

$$q = \alpha_2(\theta_2 - \theta_b) \tag{5.10}$$

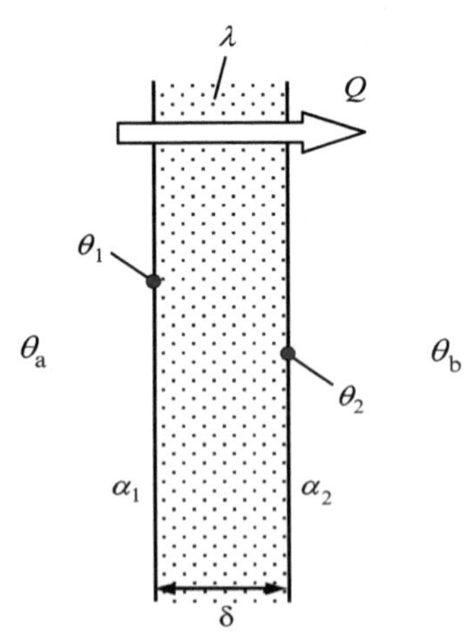

Fig. 5.1 One-dimensional heat conduction and convection system

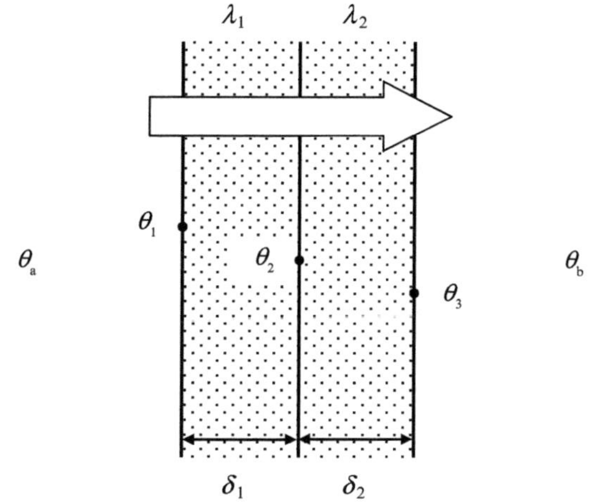

Fig. 5.2 One-dimensional heat transfer through two plates

The heat flux in the plate is expressed by

$$q = \frac{\lambda}{\delta}(\theta_1 - \theta_2) \tag{5.11}$$

Equations (5.9), (5.10), and (5.11) can be rewritten respectively as

$$\theta_c - \theta_1 = \frac{q}{\alpha_1} \tag{5.12}$$

$$\theta_1 - \theta_2 = \frac{\delta}{\lambda} q \tag{5.13}$$

$$\theta_2 - \theta_b = \frac{q}{\alpha_2} \tag{5.14}$$

Addition of these equations gives

$$\theta_a - \theta_b = \left(\frac{1}{\alpha_1} + \frac{\delta}{\lambda} + \frac{1}{\alpha_2}\right) q \tag{5.15}$$

The heat flux, q, can be derived from Eq. (5.15), thus:

$$q = \frac{\theta_a - \theta_b}{\left(\frac{1}{\alpha_1} + \frac{\delta}{\lambda} + \frac{1}{\alpha_2}\right)} \tag{5.16}$$

If a plate (thickness: δ_1, thermal conductivity: λ_1) is covered with an insulator (thickness: δ_2, thermal conductivity: λ_2) as shown in Fig. 5.2, the heat flux is given by

$$q = \frac{\theta_a - \theta_b}{\left(\frac{1}{\alpha_1} + \frac{\delta_1}{\lambda_1} + \frac{\delta_2}{\lambda_2} + \frac{1}{\alpha_2}\right)} \tag{5.17}$$

From a practical point of view, it is convenient to express Eq. (5.17) as

$$q = K(\theta_a - \theta_b) \tag{5.18}$$

where K is the overall heat-transfer coefficient:

$$K = \frac{1}{\left(\frac{1}{\alpha_1} + \frac{\delta_1}{\lambda_1} + \frac{\delta_2}{\lambda_2} + \frac{1}{\alpha_2}\right)} \tag{5.19}$$

5.3 Heat Transfer Through Multiple Plates

5.3.1 Steady-State Conduction

When plates are fixed, as shown in Fig. 5.3, the heat flux is given by

$$q = \frac{(\theta_1 - \theta_{n+1})}{\Sigma\left(\frac{\delta_i}{\lambda_i}\right)} \tag{5.20}$$

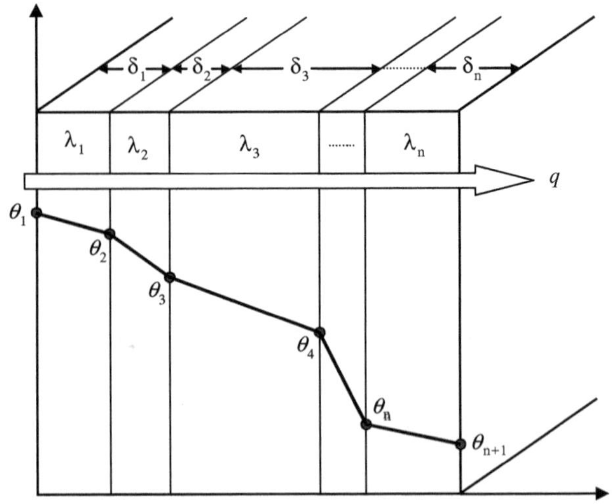

Fig. 5.3 One-dimensional heat conduction through n-plates

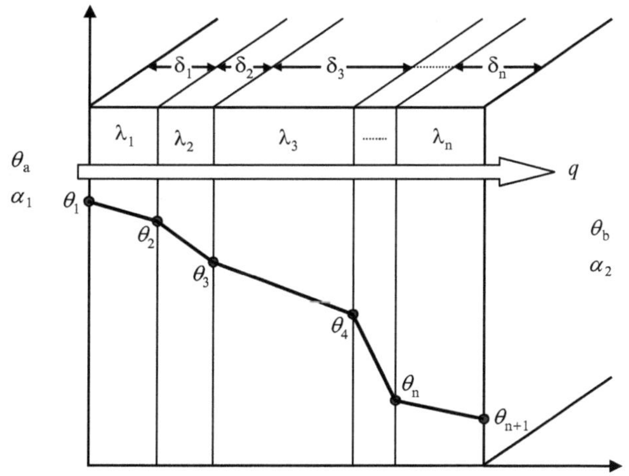

Fig. 5.4 One-dimensional heat transfer through n-plates

5.3.2 Combined Convection and Conduction

Equation (5.20) can be modified to give the following relation in the presence both of convection and conduction (see Fig. 5.4):

$$q = \frac{(\theta_a - \theta_b)}{\left[\frac{1}{\alpha_1} + \sum\left(\frac{\delta_i}{\lambda_i}\right) + \frac{1}{\alpha_2}\right]} \quad (5.21)$$

Exercise 5.1 Derive Eq. (5.21).

5.4 Conduction Through a Long Hollow Cylinder

5.4.1 Steady-State Conduction

Figure 5.5 shows a part of an infinitely long hollow cylinder subjected to conduction in the radial direction. Fourier's law yields

$$Q = -\lambda \frac{d\theta}{dr} At \quad (5.22)$$

where the area A is not constant but increases with an increase in the radial distance, r; thus,

$$A = 2\pi rl \quad (5.23)$$

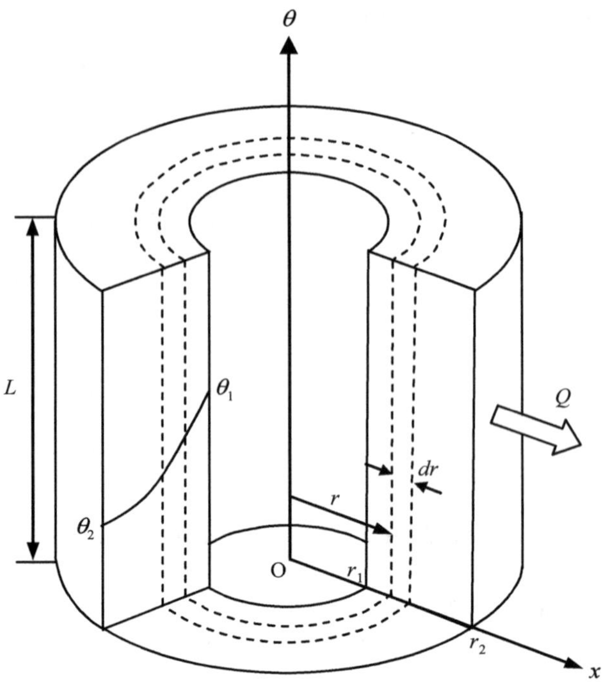

Fig. 5.5 Heat conduction through a hollow cylinder

in which l is the height of the cylinder. The introduction of the concept of heat flux is not appropriate in this case. Substitution of Eq. (5.23) into Eq. (5.22) gives

$$Q = -2\pi r l \lambda \frac{d\theta}{dr} t \tag{5.24}$$

The heat transferred by conduction per unit time is denoted by $\dot{Q}$:

$$\dot{Q} = \frac{Q}{t} \tag{5.25}$$

This $\dot{Q}$ is assumed constant. Substituting Eq. (5.24) into Eq. (5.25) gives

$$\dot{Q} = -2\pi l \lambda r \frac{d\theta}{dr} \tag{5.26}$$

Equation (5.26) is transformed into

$$d\theta = -\frac{\dot{Q}}{2\pi l \lambda} \cdot \frac{dr}{r} \tag{5.27}$$

This ordinary differential equation is solved subject to the following boundary conditions:

$$\theta = \theta_1 \quad r = r_1 \tag{5.28}$$

$$\theta = \theta_2 \quad r = r_2 \tag{5.29}$$

The following solution can be obtained by integrating Eq. (5.27):

$$\theta = -\frac{\dot{Q}}{2\pi l\lambda}\ln r + C \tag{5.30}$$

where C is the integration constant which can be determined by making use of the boundary condition, Eq. (5.28), as follows:

$$\theta_1 = -\frac{\dot{Q}}{2\pi l\lambda}\ln r_1 + C \tag{5.31}$$

$$C = \theta_1 + \frac{\dot{Q}}{2\pi l\lambda}\ln r_1 \tag{5.32}$$

Combination of Eqs. (5.32) and (5.30) gives

$$\theta = -\frac{\dot{Q}}{2\pi l\lambda}\ln r + \frac{\dot{Q}}{2\pi l\lambda}\ln r_1 + \theta_1 \tag{5.33}$$

Equation (5.33) results in the following equation subject to the boundary condition of Eq. (5.29):

$$\theta_2 = -\frac{\dot{Q}}{2\pi l\lambda}\ln r_2 + \frac{\dot{Q}}{2\pi l\lambda}\ln r_1 + \theta_1 \tag{5.34}$$

$$\theta_1 - \theta_2 = \frac{\dot{Q}}{2\pi l\lambda}(\ln r_2 - \ln r_1) \tag{5.35}$$

$$\ln r_2 - \ln r_1 = \ln\left(\frac{r_2}{r_1}\right) \tag{5.36}$$

Combination of Eqs. (5.36) and (5.35) yields

$$\theta_1 - \theta_2 = \frac{\dot{Q}}{2\pi l\lambda}\ln\left(\frac{r_2}{r_1}\right) \tag{5.37}$$

The heat transferred by conduction per unit time is expressed by

$$\dot{Q} = \frac{2\pi l\lambda}{\ln\left(\frac{r_2}{r_1}\right)}(\theta_1 - \theta_2) \tag{5.38}$$

If the inner and outer diameters, d_1 and d_2, are used in place of the radii, r_1 and r_2, Eq. (5.38) is rewritten:

$$\dot{Q} = \frac{2\pi l\lambda}{\ln\left(\frac{d_2}{d_1}\right)}(\theta_1 - \theta_2) \tag{5.39}$$

The heat transferred per unit time and unit length is expressed by

$$\frac{\dot{Q}}{l} = \frac{2\pi\lambda}{\ln\left(\frac{d_2}{d_1}\right)}(\theta_1 - \theta_2) \tag{5.40}$$

5.4.2 *Combined Conduction and Convection*

When conduction and convection heat transfer occur simultaneously (see Fig. 5.6), the heat transferred per unit time can be expressed by

$$\dot{Q} = \frac{2\pi l\lambda(\theta_a - \theta_b)}{\frac{\lambda}{\alpha_1 r_1} + \ln\frac{r_2}{r_1} + \frac{\lambda}{\alpha_2 r_2}} \tag{5.41}$$

5.5 Heat Transfer Through Composite Hollow Cylinder

5.5.1 *Steady-State Conduction*

If a vertical hollow cylinder is covered with insulator (see Fig. 5.7), Eq. (5.37) can be applied for the cylinder:

$$\theta_1 - \theta_2 = \frac{\dot{Q}}{2\pi l\lambda_1}\ln\left(\frac{r_2}{r_1}\right) \tag{5.42}$$

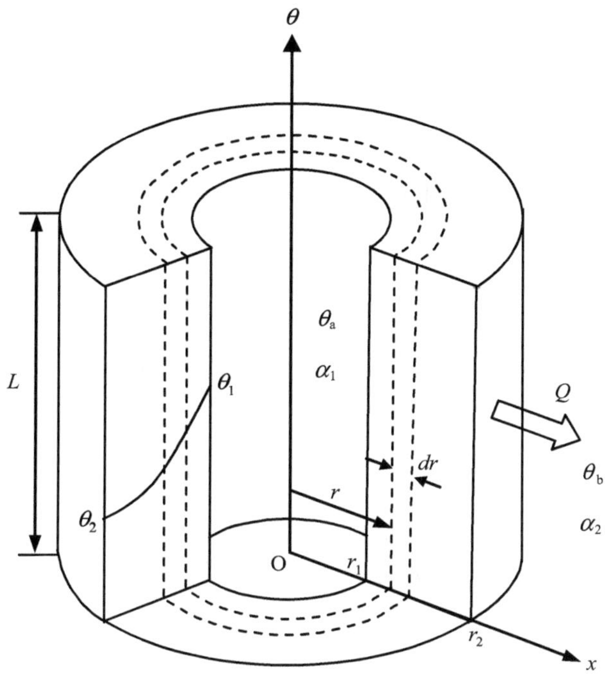

Fig. 5.6 Heat transfer through a hollow cylinder

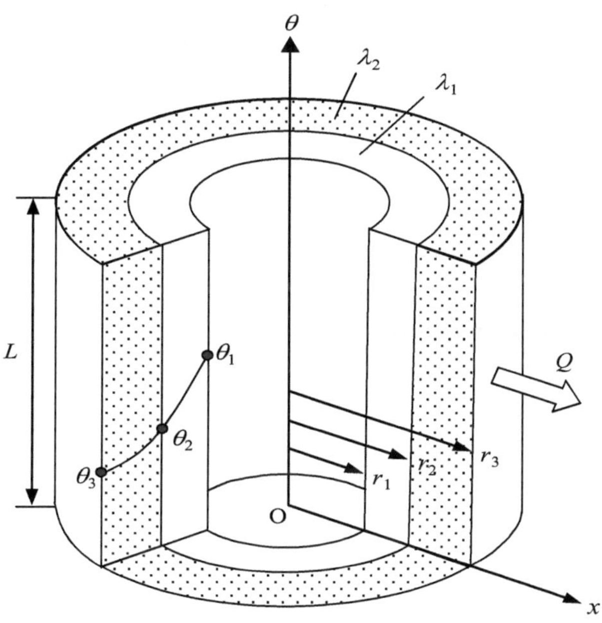

Fig. 5.7 Heat conduction through a composite cylinder

The following equation holds for the insulator:

$$\theta_2 - \theta_3 = \frac{\dot{Q}}{2\pi l \lambda_2} \ln\left(\frac{r_3}{r_2}\right) \tag{5.43}$$

Addition of Eqs. (5.42) and (5.43) yields

$$\theta_1 - \theta_3 = \frac{\dot{Q}}{2\pi l}\left\{\frac{1}{\lambda_1} \ln\left(\frac{r_2}{r_1}\right) + \frac{1}{\lambda_2} \ln\left(\frac{r_3}{r_2}\right)\right\} \tag{5.44}$$

The heat thus transferred is expressed by

$$\dot{Q} = \frac{2\pi l(\theta_1 - \theta_3)}{\left\{\frac{1}{\lambda_1} \ln\left(\frac{r_2}{r_1}\right) + \frac{1}{\lambda_2} \ln\left(\frac{r_3}{r_2}\right)\right\}} \tag{5.45}$$

The following equation is valid for multiple-layer cylindrical walls:

$$\dot{Q} = \frac{2\pi l(\theta_1 - \theta_{n+1})}{\sum_{i=1}^{n} \frac{1}{\lambda_i} \ln \frac{r_{i+1}}{r_i}} \tag{5.46}$$

5.5.2 *Combined Convection and Conduction*

$$\dot{Q} = \frac{2\pi l(\theta_a - \theta_b)}{\frac{1}{\alpha_1 r_1} + \sum_{i=1}^{n} \frac{1}{\lambda_i} \ln \frac{r_{i+1}}{r_i} + \frac{1}{\alpha_2 r_{n+1}}} \tag{5.47}$$

where θ_a and θ_b are the temperatures at location a and location b, respectively.

Exercise 5.2 Derive Eq. (5.47).

Chapter 6
Convection Heat Transfer

Abstract Natural and forced convection heat transfers are treated in this chapter. Emphasis is placed on the heat-transfer phenomena around a sphere, a cylinder, and a flat plate because of their practical importance.

Keywords Forced convection • Grashof number • Natural convection • Newton's law of cooling • Nusselt number • Prandtl number • Reynolds number

6.1 Convection from Sphere and Circular Cylinder

Convection due to temperature difference between a solid body and surrounding fluid is called natural convection. Some examples of this type of convection are shown in Figs. 6.1 and 6.2. The convection occurring when a solid body is immersed in a flowing fluid is called forced convection (see Figs. 6.3 and 6.4). In this chapter the temperature on the surface of the solid body is assumed constant.

6.1.1 Natural Convection

This section discusses natural convection from a sphere (see Fig. 6.1) and a horizontal circular cylinder (see Fig. 6.2). The heat-transfer coefficients between the solid and the flowing fluid are given below.

M. Iguchi and O.J. Ilegbusi, *Basic Transport Phenomena in Materials Engineering*, DOI 10.1007/978-4-431-54020-5_6,

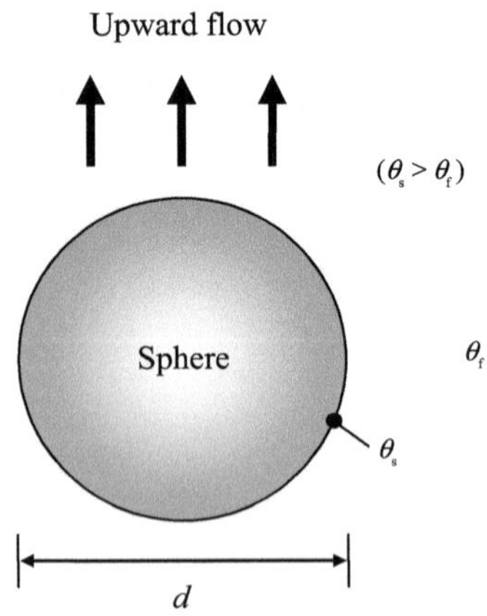

Fig. 6.1 Natural convection from a sphere

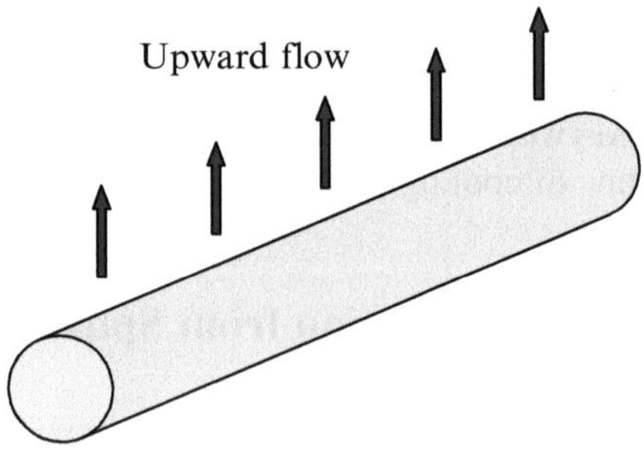

Fig. 6.2 Natural convection from a horizontal circular cylinder

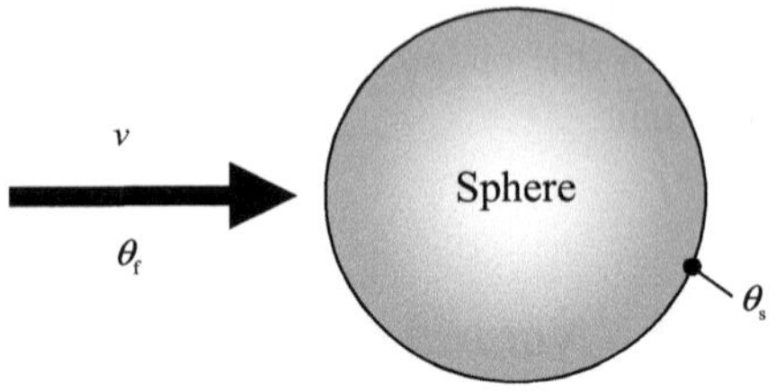

Fig. 6.3 Forced convection from a sphere

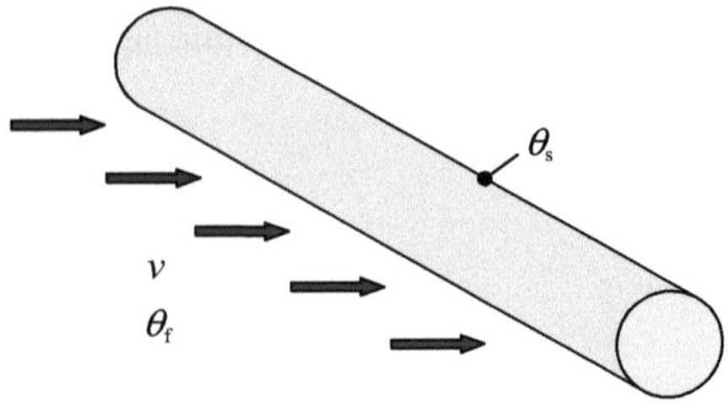

Fig. 6.4 Forced convection from a horizontal circular cylinder

6.1.1.1 Sphere

Ranz and Marshall [1] proposed the following correlation:

$$Nu = 2 + 0.60Pr^{1/3} \cdot Gr^{1/4}$$
$$(0.6 < Pr < 380, 0.9 < Pr^{1/3} \cdot Gr^{1/4} < 150) \tag{6.1}$$

where

$$Nu : \text{Nusselt number}\left(Nu = \frac{\alpha d}{\lambda_{\mathrm{f}}}\right) \tag{6.2}$$

$$Pr : \text{Prandtl number}\left(Pr = \frac{\nu_f}{\kappa_{\mathrm{f}}}\right) \tag{6.3}$$

$$Gr : \text{Grashof number}\left(Gr = \frac{g\beta d^3(\theta_{\mathrm{s}} - \theta_{\mathrm{f}})}{{\nu_f}^2}\right) \tag{6.4}$$

In the above relations, the symbols are as follows: β is the volume coefficient of expansion ($1/\theta_{\mathrm{f}}$), θ_{f} is the temperature of fluid far from the sphere (K), ν_{f} is the kinematic viscosity ($\mathrm{m^2/s}$), λ_{f} is the thermal conductivity (W/m K), κ_{f} is the thermal diffusivity ($\mathrm{m^2/s}$), g is the acceleration due to gravity ($\mathrm{m/s^2}$), d is the sphere diameter (m), θ_{s} is the surface temperature of solid (K), θ_{fm} is the mean temperature of fluid (K).

$$\theta_{\mathrm{fm}} = \frac{\theta_{\mathrm{s}} + \theta_{\mathrm{f}}}{2} \tag{6.5}$$

The physical properties of a fluid with the exception of β are typically evaluated at the temperature θ_{f}. Thus,

$$Pr = Pr(\theta_{\mathrm{fm}}), \nu_{\mathrm{f}} = \nu_{\mathrm{f}}(\theta_{\mathrm{fm}}), \lambda_{\mathrm{f}} = \lambda_{\mathrm{f}}(\theta_{\mathrm{fm}}), \beta = \beta(\theta_{\mathrm{f}}) \tag{6.6}$$

6.1.1.2 Horizontal Circular Cylinder

McAdams [1, 2] derived the following equation for the Nusselt number as a function of the Grashof number (Gr), and the Prandtl number (Pr), for horizontal cylinder of unit length:

$$Nu = 0.53(Gr \cdot Pr)^{1/4}$$
$$(10^4 < Gr \cdot Pr < 10^8) \tag{6.7}$$

6.1.1.3 Meaning of Dimensionless Numbers

The Nusselt number (Nu) can be defined as the ratio of heat convection to heat conduction thus:

$$\frac{Q_{\text{conv}}}{Q_{\text{cond}}} = \frac{\alpha(\theta_s - \theta_f)At}{\lambda_f \frac{\theta_s - \theta_f}{d} At} = \frac{\alpha d}{\lambda_f} \tag{6.8}$$

The Prandtl number (Pr) is a measure of the ratio of the hydrodynamic boundary layer thickness δ_h to the thermal boundary layer thickness δ_{th}: $Pr = (\delta_h/\delta_{th})^2$ in the case of Stokes' first problem.

The Grashof number (Gr) expresses the product of the Reynolds number and the ratio of buoyancy force to viscous force, thus:

$$\begin{aligned} Gr &= \frac{g\beta d^3(\theta_s - \theta_f)}{{\nu_f}^2} \quad (\nu_f:\ \text{Kinematic viscosity}) \\ &= \frac{\nu d}{\nu_f} \cdot \frac{\rho g d^3 \beta(\theta_s - \theta_f)}{\rho \nu_f \frac{\nu}{d} d^2} \end{aligned} \tag{6.9a}$$

$$= \text{Reynolds number} \times \frac{\text{Buoyancy force}}{\text{Viscous force}} \tag{6.9b}$$

Exercise 6.1 Determine the heat-transfer coefficient for a sphere of diameter 8.0 cm placed in air ambience at 27 °C. The temperature of the sphere is 227 °C. Use Table PII.1 in Appendix to evaluate the physical properties of air. (Answer: 11.2 W/m^2 K)

Exercise 6.2 A long circular cylinder of 20 cm diameter is placed horizontally in the atmosphere. The temperatures of the cylinder and air are 100 °C and 20 °C, respectively. Determine the heat transferred from the cylinder per unit time (1 s) and unit length (1 m). The physical properties of air at a mean temperature of 60 °C are as follows: $Pr = 0.71$, $\nu_f = 0.196 \times 10^{-4}$ m^2/s, and $\lambda_f = 0.0287$ W/m K. (Answer: 303 W/m)

Answer to Exercise 6.1 The physical properties of fluid should be determined at the mean temperature between the temperatures of the solid and fluid, θ_{fm}.

$$\theta_{fm} = (\theta_s + \theta_f)/2 = (227 + 27)/2 = 127\,^\circ\text{C} = 400\,\text{K} \tag{6.10}$$

The following values can be obtained from Table PII.1:

$$\nu_f = 26.39\,\text{mm}^2/\text{s} = 26.39 \times 10^{-6}\,\text{m}^2/\text{s} \tag{6.11}$$

$$Pr = 0.715 \tag{6.12}$$

$$\lambda_f = 33.05 \times 10^{-3} \mathrm{W/m\,K} \tag{6.13}$$

The volume coefficient of expansion, β, is determined at the fluid temperature, θ_f:

$$\theta_f = 27 + 273 = 300\,\mathrm{K} \tag{6.14}$$

The definition of β yields

$$\beta = 1/\theta_f = 1/300 \tag{6.15}$$

The Grashof number is calculated by substituting the following quantities into Eq. (6.4):

$$d = 0.08\,\mathrm{m} \tag{6.16}$$

$$(\theta_s - \theta_f) = 227 - 27 = 200\,\mathrm{K} \tag{6.17}$$

$$g = 9.80\,\mathrm{m/s^2} \tag{6.18}$$

$$\begin{aligned} Gr &= 9.8 \times (1/300) \times (0.08)^3 \times 200 \times \left[1/(26.39 \times 10^{-6})^2\right] \\ &= 4.803 \times 10^6 \end{aligned} \tag{6.19}$$

From Eqs. (6.12) and (6.19) we have

$$Pr^{1/3} \cdot Gr^{1/4} = (0.715)^{1/3} \times (4.803 \times 10^6)^{1/4} = 41.86 \tag{6.20}$$

This value is included in the applicable range of Eq. (6.1). Thus, the Nusselt number is calculated from Eq. (6.1).

$$\begin{aligned} Nu &= 2 + 0.60 Pr^{1/3} \cdot Gr^{1/4} = 2 + 0.60(0.715)^{1/3} \times (4.803 \times 10^6)^{1/4} \\ &= 2 + 0.60 \times 41.86 = 27.12 \end{aligned} \tag{6.21}$$

The heat-transfer coefficient, α, is determined from the definition of the Nusselt number:

$$\begin{aligned} \alpha &= 27.12 \lambda_f / d = 27.12 \times (33.05 \times 10^{-3})/0.08 \\ &= 11.2\,\mathrm{W/m^2\,K} \end{aligned} \tag{6.22}$$

Answer to Exercise 6.2 The Grashof number (Gr), can be calculated from Eq. (6.4) by substituting the following quantities:

$$\theta_f = 20 + 273 = 293\,\mathrm{K} \tag{6.23}$$

$$(\theta_s - \theta_f) = 100 - 20 = 80\,\mathrm{K} \tag{6.24}$$

$$d = 0.20\,\text{m} \tag{6.25}$$

$$\beta = 1/\theta_f = 1/293 \tag{6.26}$$

$$g = 9.80\,\text{m/s}^2 \tag{6.27}$$

Equation (6.4) gives

$$\begin{aligned} Gr &= 9.8 \times (1/293) \times (0.20)^3 \times 80 \times \left[1/(0.196 \times 10^{-4})^2\right] \\ &= 5.572 \times 10^7 \end{aligned} \tag{6.28}$$

$$Pr \cdot Gr = 0.71 \times 5.572 \times 10^7 = 3.956 \times 10^7 \tag{6.29}$$

This value is included in the applicable range of Eq. (6.7). The Nusselt number is therefore

$$\begin{aligned} Nu &= 0.53(Gr \cdot Pr)^{1/4} = 0.53 \times (3.956 \times 10^7)^{1/4} \\ &= 42.03 \end{aligned} \tag{6.30}$$

The definition of the Nusselt number gives

$$\begin{aligned} \alpha &= 42.03\lambda_f/d = 42.03 \times (0.0287)/0.20 \\ &= 6.03\,\text{W/m}^2\,\text{K} \end{aligned} \tag{6.31}$$

The heat transferred by convection in 1 s is given by

$$Q = \alpha A(\theta_s - \theta_f) \tag{6.32}$$

where the surface area, A,

$$A = \pi dl = 3.14 \times 0.20 \times 1.00 = 0.6283\,\text{m}^2 \tag{6.33}$$

Substitution of Eqs. (6.24), (6.31), and (6.33) into Eq. (6.32) yields

$$Q = 6.03 \times 0.6283 \times 80 = 303\,\text{W/m} \tag{6.34}$$

6.1.2 *Forced Convection*

6.1.2.1 Sphere (See Fig. 6.3)

Ranz and Marshall [1, 3] derived the following equation for the Nusselt number (Nu), of a sphere placed in a flow with approach velocity, v:

$$Nu = 2 + 0.60Pr^{1/3} \cdot Re^{1/2}$$
$$(0.6 < Pr < 380, 1 < Re < 10^5) \tag{6.35}$$

$$Nu = \frac{\alpha d}{\lambda_f}, \; Pr = \frac{\nu_f}{\kappa_f}, \; Re = \frac{v d}{\nu_f} \tag{6.36}$$

6.1.2.2 Circular Cylinder (See Fig. 6.4)

Eckert [1, 3] proposed a relation of the form

$$Nu = 1.11CPr^{0.31} \cdot Re^n \tag{6.37}$$

The coefficient, C, and the index, n, are expressed as follows:

$$C = 0.891 \text{ and } n = 0.330 \text{ for } Re = 0.4 - 4 \tag{6.38}$$

$$C = 0.821 \text{ and } n = 0.385 \text{ for } Re = 4 - 40 \tag{6.39}$$

$$C = 0.615 \text{ and } n = 0.466 \text{ for } Re = 40 - 4 \times 10^3 \tag{6.40}$$

$$C = 0.174 \text{ and } n = 0.618 \text{ for } Re = 4 \times 10^3 - 4 \times 10^4 \tag{6.41}$$

$$C = 0.0239 \text{ and } n = 0.805 \text{ for } Re = 4 \times 10^4 - 4 \times 10^5 \tag{6.42}$$

Exercise 6.3 A sphere of diameter 5.0 cm is placed in air flowing at velocity 20 m/s. The temperatures of the sphere and air are 100 °C and 20 °C, respectively. Determine the heat-transfer coefficient for the sphere, α, and the heat transferred per unit time, Q. The physical properties of air at a mean temperature of 60 °C are $Pr = 0.71$, $\nu_f = 0.196 \times 10^{-4}$ m²/s, and $\lambda_f = 0.0287$ W/m K. (Answer: 70.5 W/m² K)

Exercise 6.4 A circular cylinder of diameter of 5.0 cm is placed in an air flow of approach velocity of 22 m/s. The temperatures of the cylinder and air are 100 °C and 20 °C, respectively. Determine the heat-transfer coefficient for the cylinder, α, and the heat transferred per unit time and unit length, Q. The physical properties of air at a mean temperature of 60 °C are $Pr = 0.71$, $\nu_f = 0.196 \times 10^{-4}$ m²/s, and $\lambda_f = 0.0287$ W/m K. (Answer: 91 W/m² K)

Answer to Exercise 6.3 The physical properties of fluid at a mean temperature $\theta_{fm} = 60$ °C are given by

$$\nu_f = 0.196 \times 10^{-4} \text{ m}^2/\text{s} \tag{6.43}$$

$$Pr = 0.71 \tag{6.44}$$

$$\lambda_f = 0.0287\,\mathrm{W/m\,K} \tag{6.45}$$

The Reynolds number can be calculated as follows:

$$d = 0.050\,\mathrm{m} \tag{6.46}$$

$$v = 20\,\mathrm{m/s} \tag{6.47}$$

$$\begin{aligned} Re &= vd/\nu_f = 20 \times 0.050/(0.196 \times 10^{-4}) \\ &= 5.10 \times 10^4 \end{aligned} \tag{6.48}$$

where both the Prandtl number and the Reynolds number are included in the applicable range of Eq. (6.35). The Nusselt number can be calculated from Eq. (6.35) as follows:

$$\begin{aligned} Nu &= 2 + 0.60Pr^{1/3} \cdot Re^{1/2} = 2 + 0.60(0.71)^{1/3} \times (5.10 \times 10^4)^{1/2} \\ &= 2 + 0.60 \times 0.8921 \times 225.8 = 2 + 120.9 \\ &= 122.9 \end{aligned} \tag{6.49}$$

The heat-transfer coefficient can be calculated in the following manner:

$$\begin{aligned} \alpha &= 122.9\,\lambda_f/d = 122.9 \times (0.0287)/0.050 \\ &= 70.5\,\mathrm{W/m^2\,K} \end{aligned} \tag{6.50}$$

The heat transferred per unit time, Q, is given by

$$\begin{aligned} Q &= \alpha(\theta_s - \theta_f)A = \alpha(\theta_s - \theta_f)\pi d^2 \\ &= 70.5 \times (100 - 20) \times \pi \times (0.05)^2 = 44.3\,\mathrm{W} \end{aligned} \tag{6.51}$$

Answer to Exercise 6.4 The physical properties at $\theta_{fm} = 60\,°\mathrm{C}$ are given by

$$\nu_f = 0.196 \times 10^{-4}\ \mathrm{m^2/s} \tag{6.52}$$

$$Pr = 0.71 \tag{6.53}$$

$$\lambda_f = 0.0287\ \mathrm{W/m\,K} \tag{6.54}$$

The Reynolds number is calculated as follows:

$$d = 0.05\,\mathrm{m} \tag{6.55}$$

$$v = 22\,\mathrm{m/s} \tag{6.56}$$

$$\begin{aligned} Re &= vd/\nu_{\mathrm{f}} = 22 \times 0.05/0.196 \times 10^{-4} \\ &= 5.61 \times 10^4 \end{aligned} \tag{6.57}$$

From Eq. (6.42) the coefficient and index of Eq. (6.37), C and n, are estimated to be 0.0239 and 0.805, respectively. Substituting these values into Eq. (6.37) yields

$$\begin{aligned} Nu &= 1.11 \times 0.0239 \times (0.71)^{0.31} \times (5.61 \times 10^4)^{0.805} \\ &= 1.11 \times 0.0239 \times 0.899 \times 6{,}650 = 158.6[-] \end{aligned} \tag{6.58}$$

Accordingly, we have

$$\begin{aligned} \alpha &= 158.6\lambda_{\mathrm{f}}/d = 158.6 \times 0.0287/0.05 \\ &= 91.0\,\mathrm{W/m^2\,K} \end{aligned} \tag{6.59}$$

The heat transferred per unit time and unit length, Q, is given by

$$\begin{aligned} Q &= \alpha(\theta_{\mathrm{s}} - \theta_{\mathrm{f}})A = \alpha(\theta_{\mathrm{s}} - \theta_{\mathrm{f}})\pi dL \\ &= 91.0 \times (100 - 20) \times \pi \times (0.05) \times 1.0 = 1{,}143\,\mathrm{W/m} \end{aligned} \tag{6.60}$$

6.2 Convection from Flat Plate

6.2.1 Natural Convection from Vertical Flat Plate

Convection from a vertical flat plate (see Fig. 6.5) is represented in the following form:

$$\mathrm{Nu_m} = C(Gr \cdot Pr)^{\mathrm{m}} \tag{6.61}$$

$$Nu_{\mathrm{m}} = \alpha_{\mathrm{m}}L/\lambda_{\mathrm{f}} \tag{6.62}$$

where Nu_{m} is the mean Nusselt number averaged between $x = 0$ and $x = L$, α_{m} is the mean heat-transfer coefficient, and L is the height of the plate. The coefficient, C, and the index, m, are expressed as follows:

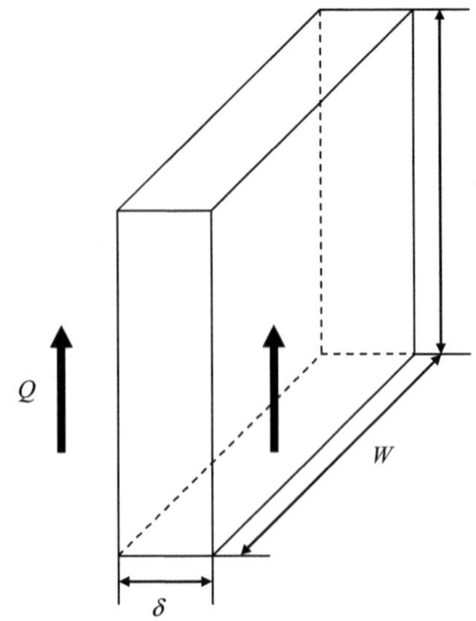

Fig. 6.5 Natural convection from a vertical flat plate

$$C = 0.59 \text{ and } m = 1/4 \text{ for } Gr \cdot Pr = 10^4 \sim 10^9 \quad (6.63)$$

$$C = 0.021 \text{ and } m = 2/5 \text{ for } Gr \cdot Pr = 10^9 \sim 10^{13} \quad (6.64)$$

The following equation is sometimes used in place of Eq. (6.64):

$$C = 0.10 \text{ and } m = 1/3 \text{ for } Gr \cdot Pr = 10^9 \sim 10^{13} \quad (6.65)$$

6.2.2 Forced Convection from Vertical Flat Plate

The Nusselt number in this case (see Fig. 6.6) can be approximated by the equation of Nu_{m} proposed for a horizontal flat plate.

6.2.3 Forced Convection from Horizontal Flat Plate

Consider the case that a flat plate is placed horizontally, as shown in Fig. 6.7).

Fig. 6.6 Forced convection from a vertical flat plate

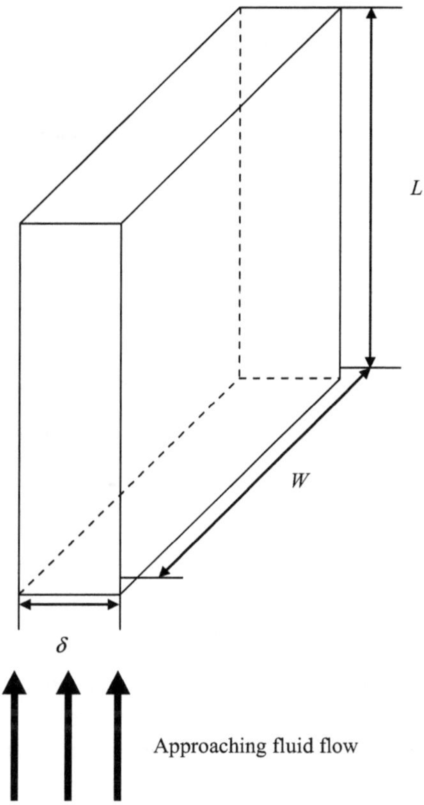

6.2.3.1 Laminar Flow

$$Nu_x = 0.332 Pr^{1/3} {Re_x}^{1/2} \quad (0.6 < Pr) \tag{6.66}$$

$$Nu_\mathrm{m} = 0.664 Pr^{1/3} {Re_\mathrm{L}}^{1/2} \quad (0.6 < Pr) \tag{6.67}$$

$$Nu_x = \alpha x / \lambda_\mathrm{f} \tag{6.68}$$

$$Nu_\mathrm{m} = \alpha_\mathrm{m} L / \lambda_\mathrm{f} \tag{6.69}$$

$$Re_x = v x / \nu_\mathrm{f} \tag{6.70}$$

$$Re_\mathrm{L} = v L / \nu_\mathrm{f} \tag{6.71}$$

where Nu_x is the local Nusselt number, Nu_m is the mean Nusselt number averaged between $x = 0$ and $x = L$, Re_x is the local Reynolds number, and Re_L is the Reynolds number based on the length of the plate, L. The physical properties of the fluid are evaluated at the film temperature, θ_fm.

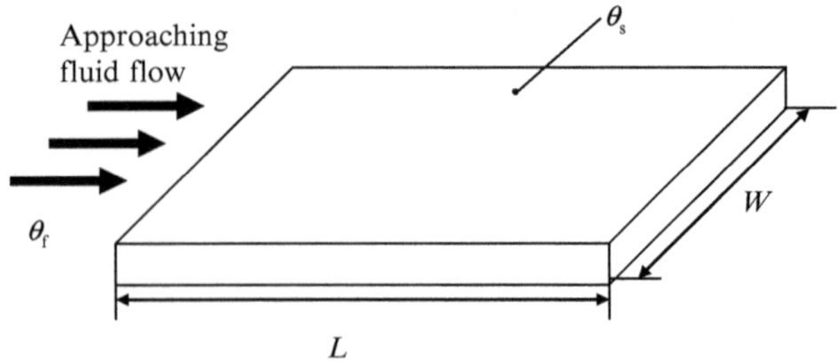

Fig. 6.7 Forced convection from a horizontal flat plate (laminar flow)

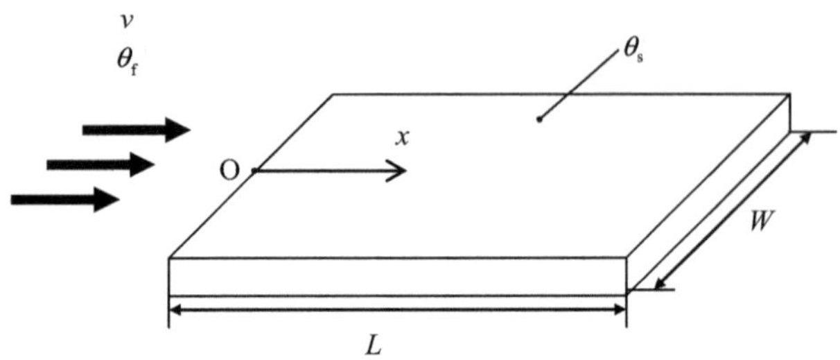

Fig. 6.8 Forced convection from a horizontal flat plate

Exercise 6.5 A flat plate of width $W = 1.0$ m and length $L = 3.0$ m is placed in parallel to air flow direction (see Fig. 6.8). The approach velocity, v, is 5.0 m/s, and the boundary layer on the plate is laminar. The temperatures of the plate and air are 100 °C and 20 °C, respectively. Calculate the local heat flux at $x = 0.80$ m, where x is measured from the leading edge of the plate. The physical properties of air at the film temperature of 60 °C are $\nu_f = 1.96 \times 10^{-4}$ m^2/s, $Pr = 0.71$, and $\lambda_f = 28.42 \times 10^{-3}$ W/m K.

Answer to Exercise 6.5 The local Reynolds number is given by

$$Re_x = vx/\nu_f = 5.0 \times 0.8/(1.96 \times 10^{-5}) = 2.04 \times 10^5 \tag{6.72}$$

This value is less than the local critical Reynolds number for a flat plate of approximately 3×10^5, and the flow at this location is laminar. The local Nusselt number, Nu_x, is calculated from Eq. (6.66) as

$$\begin{aligned} Nu_x &= 0.332 Pr^{1/3} {Re_x}^{1/2} = 0.332 \times (0.71)^{1/3} \times (2.04 \times 10^5)^{1/2} \\ &= 0.332 \times 0.892 \times 452 = 134 \end{aligned} \tag{6.73}$$

$$\alpha_x = \lambda_f Nu_x/x = 28.42 \times 10^{-3} \times 134/0.80 = 4.76\,\mathrm{W/m^2\,K} \tag{6.74}$$

$$q = \alpha_x(\theta_s - \theta_f) = 4.76 \times (100 - 20) = 4.76 \times 80 = 381\,\mathrm{W} \tag{6.75}$$

6.2.3.2 Turbulent Flow

$$Nu_x = 0.0296Re_x^{\;0.8}Pr/\left[1 + BRe_x^{\;-0.1}(Pr - 1)\right] \; (0.5 < Pr < 5) \tag{6.76}$$

$$B = 0.860\,[1 + \ln\{(1 + 5Pr)/6\}]/(Pr - 1) \tag{6.77}$$

$$Nu_\mathrm{m} = 0.037Re_\mathrm{L}^{4/5}Pr^{1/3} \quad (Pr \approx 1) \tag{6.78}$$

6.3 Phase Transformation

This section describes convection processes associated with a change in the phase of the fluid, specifically solidification and boiling heat transfer. Heat-transfer rates are highly elevated under phase change conditions.

6.3.1 Solidification and Melting

Heat transfer occurs between metal and the surroundings during solidification and fusion (melting) of metal. The melting temperature, T_m, and latent heat of fusion, ΔH_m, of metals extensively used in practical applications are given in Table 6.1 below.

6.3.2 Boiling Heat Transfer

Boiling heat transfer occurs between a liquid and the surface of a solid body which is heated in the liquid above the saturation temperature, θ_sat. When the surface is immersed in a liquid pool without any liquid motion (see Fig. 6.9), the process is referred to as pool boiling. On the other hand, if the surface is immersed in flowing

Table 6.1 Melting temperature and latent heat of fusion of common metals

Material	T_m (K)	ΔH_m (KJ/Kg)
Aluminum	933.5	395
Bismuth	544.6	50.2
Mercury	234.28	11.7
Sodium	371	115
Lead	600.7	24.7
Tin	505.1	60.6
Zinc	692.7	102
Iron	1,810	267
Copper	1,357.6	205

Fig. 6.9 Pool boiling heat transfer

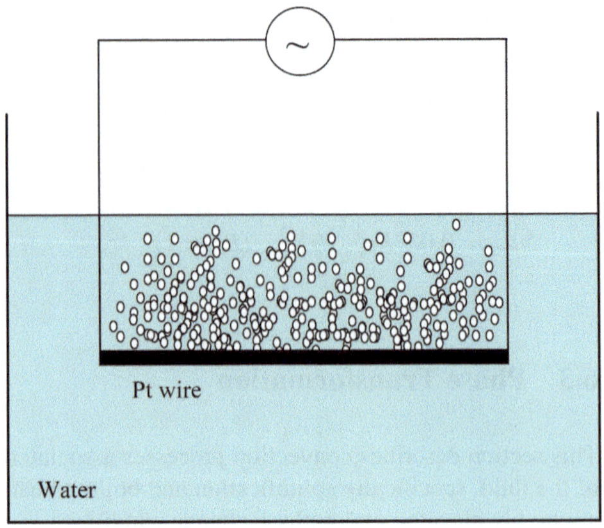

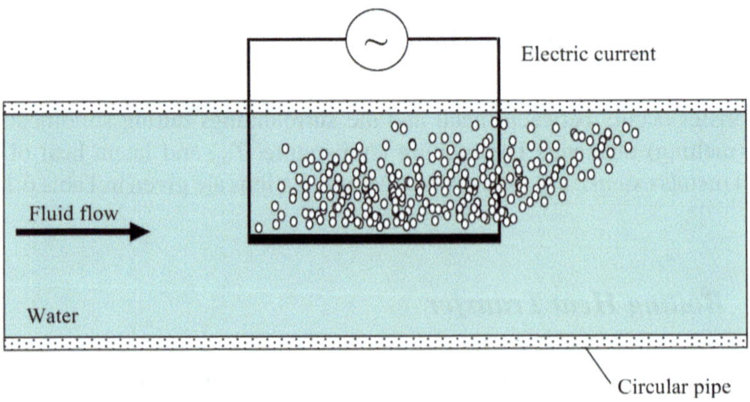

Fig. 6.10 Forced convection boiling

liquids such as a liquid pipe flow (see Fig. 6.10), the process is called forced convection boiling. Only pool boiling is introduced in this section.

6.3.2.1 Classification of Pool Boiling Heat Transfer

Some heat-transfer patterns appear depending on the difference in temperature between the surface, θ_w, and the saturation temperature, θ_{sat}, as schematically shown in Fig. 6.11.

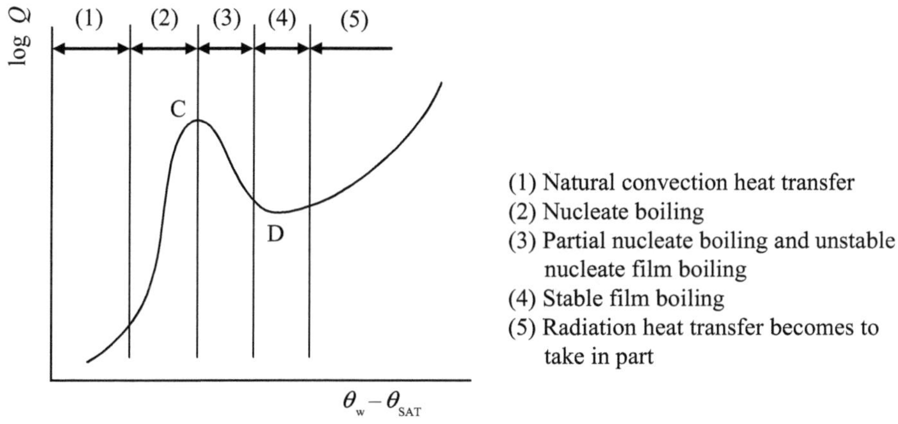

Fig. 6.11 Classification of boiling heat transfer

1. *Natural convection heat transfer*: The liquid near the surface evaporates. The heat-transfer rate increases slightly with increase in the temperature difference, $\theta_w - \theta_{sat}$.
2. *Nucleate boiling*: Many bubbles are generated on the surface and dissipate in the liquid after leaving the surface. The heat-transfer rate increases significantly with temperature difference, $\theta_w - \theta_{sat}$.
3. *Partial nucleate boiling and unstable nucleate film boiling*: Vapor films partly cover the surface. These films suppress the heat transfer, and, hence, the heat-transfer rate decreases as the covered surface area increases.
4. *Stable film boiling*: Stable vapor films are generated on the surface. The heat-transfer rate is kept almost constant.
5. *Radiation heat transfer becomes important*: The heat-transfer rate increases again with $\theta_w - \theta_{sat}$.

In Fig. 6.11 point C is called the burnout point or the maximum heat flux point, and point D is the minimum heat flux point.

6.3.2.2 Factors Affecting Boiling Heat Transfer

1. Gases dissolved in a liquid pool
2. Liquid pressure
3. Surface roughness

6.3.2.3 Empirical Relation for Heat-Transfer Rate

Nucleate Boiling

Stefan and Abdelsalam [4] proposed the following empirical relation for nucleate boiling heat transfer in an arbitrary liquid:

$$Nu = 0.23{X_1}^{0.674}{X_2}^{0.35}{X_4}^{0.371}{X_5}^{0.297}{X_{13}}^{-1.73} \quad (10^{-4} \le p/p_{\mathrm{cr}} \le 0.97) \tag{6.79}$$

$$Nu = \frac{\alpha d_B}{\lambda_{\mathrm{L}}} \tag{6.80}$$

$$X_1 = \frac{q d_B}{\lambda_{\mathrm{L}} \theta_{\mathrm{sat}}} \tag{6.81}$$

$$X_2 = \frac{\rho_{\mathrm{L}} {\kappa_{\mathrm{L}}}^2}{\sigma d_B} \tag{6.82}$$

$$X_4 = \frac{\Delta h {d_B}^2}{{\kappa_L}^2} \tag{6.83}$$

$$X_5 = \frac{\rho_{\mathrm{v}}}{\rho_{\mathrm{L}}} \tag{6.84}$$

$$X_{13} = \frac{(\rho_{\mathrm{L}} - \rho_{\mathrm{v}})}{\rho_{\mathrm{L}}} \tag{6.85}$$

$$d_B = 0.0146 \theta_{\mathrm{c}} \left[\frac{2\sigma}{g(\rho_{\mathrm{L}} - \rho_{\mathrm{v}})} \right]^{1/2} \tag{6.86}$$

where p (Pa) is the pressure, p_{cr} is the critical pressure (Pa), Nu is the Nusselt number, α is the heat-transfer coefficient (W/m^2 K), d_{B} is the bubble diameter (m), κ_{L} is the thermal diffusivity of liquid (m^2/s), q is the heat flux (W/m^2), θ_{sat} is the saturation temperature (K), ρ_{L} is the density of liquid (kg/m^3), σ is the surface tension (N/m), Δh is the latent heat of evaporation (J/kg), ρ_{v} is the density of vapor (kg/m^3), θ_{c} is the contact angle of liquid (°), and g is the acceleration due to gravity (m/s^2).

Stable Film Boiling

A vapor film exists between the surface of a solid body and liquid. The surface of the solid body is completely covered with the vapor film. The surface temperature,

θ_w, is usually much higher than the saturation temperature, θ_{sat}. The Nusselt number is given by

$$Nu = C_1 \left(\frac{Gr * Pr_v}{Sp} \right) \tag{6.87}$$

$$\alpha = \frac{q_w}{(\theta_w - \theta_{sat})} \tag{6.88}$$

$$Gr* = \frac{g\rho_v(\rho_L - \rho_v)L^3}{\mu_v{}^2} \tag{6.89}$$

$$Pr_v = \frac{c_{pv}\mu_v}{\lambda_v} \tag{6.90}$$

$$Sp = \frac{c_{pv}(\theta_w - \theta_{sat})}{L'} \tag{6.91}$$

$$L' = \Delta h_v \left[1 + C_2 \frac{c_{pv}(\theta_w - \theta_{sat})}{\Delta h_v} \right] \tag{6.92}$$

where Gr^* is the Grashof number (−), Pr_v is the Prandtl number of vapor (−), α is the heat-transfer coefficient (W/m^2 K), q_w is the heat flux on the surface (W/m^2), θ_w is the surface temperature (K), θ_{sat} is the saturation temperature (K), ρ_L is the density of liquid (kg/m^3), ρ_v is the density of vapor (kg/m^3), L is the representative length (m), μ_L is the dynamic viscosity of liquid (kg/m s), c_{pv} is the specific heat of vapor at constant pressure (J/kg K), λ_v is the thermal conductivity of vapor (W/m K), Δh_v is the latent heat of evaporation (J/kg), L' is the modified latent heat of evaporation (J/kg).

The physical properties of vapor should be evaluated at the film temperature and those of the liquid phase at the saturation temperature. The representative length, L, and the coefficients, C_1 and C_2, for some solid bodies used frequently in practical applications are given as follows:

(i) Horizontal circular cylinder: $L = D$ (diameter), $C_1 = 0.62$, $C_2 = 0.68$
(ii) Sphere: $L = D$ (diameter), $C_1 = 0.67$, $C_2 = 0.68$

References

1. Japan Society of Mechanical Engineers (1980) Data book on heat transfer, 3rd edn. JSME, Tokyo
2. Holman JP (1972) Heat transfer, 3rd edn. McGraw-Hill Kogakusha, Tokyo
3. Shiba K (1973) Thermodynamics. Iwanami Book Co., Ltd., Tokyo p 10, 26
4. Stephan K, Abdelsalam M (1980) Heat-transfer correlations for natural convection boiling. Int J Heat Mass Trans 23–1:73

Chapter 7
Radiation Heat Transfer

Abstract In radiation heat transfer energy is transferred at the speed of light even in the absence of any medium between the heat source and the surface of the solid body of concern. Some relevant fundamental concepts and laws are first introduced in this chapter. Subsequently some examples are provided to illustrate the application of these laws.

Keywords Black body • Gray body • Radiation • Shape factor • Stefan–Boltzmann law

7.1 Basic Definitions

A part of radiation energy incident on the surface of a body is absorbed by the body, another part is transmitted through the body, and the remaining part is reflected from the surface of the body (see Fig. 7.1). The fraction of the incident radiation that is absorbed is called absorptivity, α^*. The transmissivity, τ^*, and reflectivity, ρ^*, are similarly defined as the fractions transmitted and reflected, respectively. When the absorptivity of a solid body is unity, the body is called a black body. A body whose absorptivity is less than unity is known as a gray body. By definition therefore,

$$\alpha * + \tau * + \rho * = 1 \tag{7.1}$$

7.2 Stefan–Boltzmann Law

The following Stefan–Boltzmann law is valid for a black body (see Fig. 7.2):

$$E_{\mathrm{B}} = \sigma_{\mathrm{B}} \left(\frac{\theta}{100} \right)^4 \tag{7.2}$$

M. Iguchi and O.J. Ilegbusi, *Basic Transport Phenomena in Materials Engineering*, DOI 10.1007/978-4-431-54020-5_7, © Springer Japan 2014

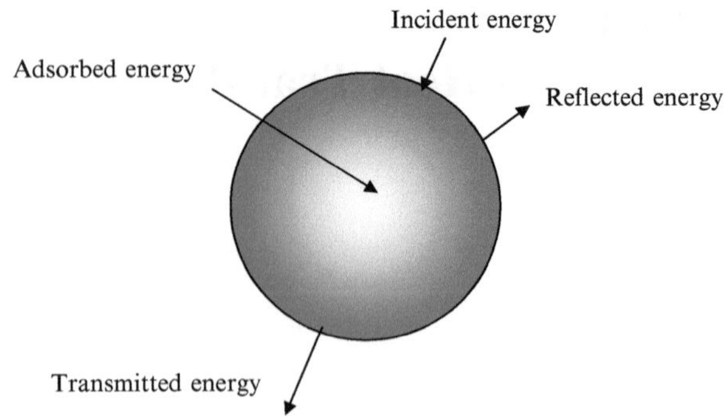

Fig. 7.1 Radiation heat transfer

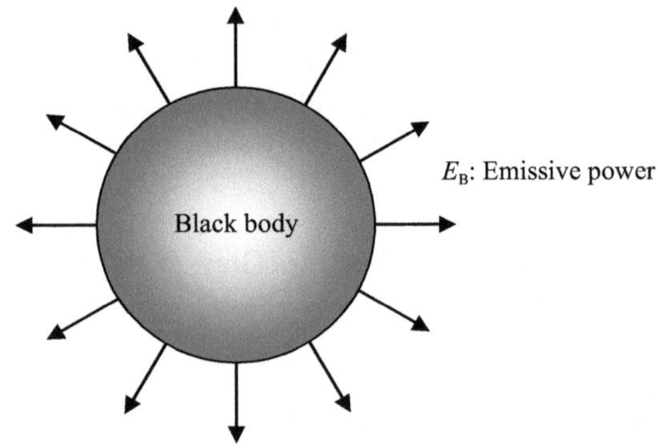

Fig. 7.2 Emissive power from a black body

where E_{B} (W/m^2) is the emissive power which denotes the radiative energy of the black body per unit area and unit time, θ is the absolute temperature, the subscript B denotes a black body, and σ_{B} is the Stefan–Boltzmann constant:

$$\sigma_{\mathrm{B}} = 5.67\,(\mathrm{W/m^2\,K^4}) \tag{7.3}$$

Exercise 7.1 The temperature of a black body is 800 K (see Fig. 7.2). Calculate the emissive power, E_{B}.

Answer to Exercise 7.1 The emissive power, E_{B}, can be calculated from Eq. (7.2) as

$$E_{\mathrm{B}} = \sigma_{\mathrm{B}}\left(\frac{\theta}{100}\right)^4 = 5.67\left(\frac{\theta}{100}\right)^4 = 5.67\left(\frac{800}{100}\right)^4 = 2.32 \times 10^4\ \mathrm{W/m^2} \tag{7.4}$$

7.3 Gray Body Radiation

The emissivity of a real solid body, $\varepsilon*$, is not unity and, hence, Eq. (7.4) is modified to give

$$E_B = \varepsilon * \sigma_B \left(\frac{\theta}{100}\right)^4 \tag{7.5}$$

According to Kirchhoff's law, the emissivity, $\varepsilon*$, is equal to the absorptivity, $\alpha*$:

$$\varepsilon* = \alpha* \tag{7.6}$$

Exercise 7.2 The temperature of a gray body is 1,000 K and the emissivity is 0.41. Calculate the emissive power, E_B.

Answer to Exercise 7.2 The emissive power can be calculated from Eq. (7.5) as

$$\begin{aligned} E_B &= \varepsilon * \sigma_B \left(\frac{\theta}{100}\right)^4 = 0.41 \times 5.67 \left(\frac{\theta}{100}\right)^4 \\ &= 2.32 \left(\frac{1,000}{100}\right)^4 = 2.32 \times 10^4 \ \mathrm{W/m^2} \end{aligned} \tag{7.7}$$

7.4 Radiation Between Two Black Bodies

The radiation energy from a black body 1 of temperature θ_1 is absorbed by a black body 2 of temperature θ_2 (see Fig. 7.3). Reverse radiation heat transfer occurs between the two bodies. The net radiation heat transfer between two black bodies, Q_{12} (W), is given by

$$\begin{aligned} Q_{12} &= \sigma_B \left[\left(\frac{\theta_1}{100}\right)^4 - \left(\frac{\theta_2}{100}\right)^4\right] A_1 F_{12} \\ &= \sigma_B \left[\left(\frac{\theta_1}{100}\right)^4 - \left(\frac{\theta_2}{100}\right)^4\right] A_2 F_{21} \end{aligned} \tag{7.8}$$

where A_1 and A_2 are the surface areas of the black bodies 1 and 2, respectively, and F_{12} and F_{21} are their respective shape factors. The shape factor or geometric factor is dependent on the surface area and shape of the two bodies in addition to their relative positions. The following relationship holds:

$$A_1 F_{12} = A_2 F_{21} \tag{7.9}$$

Some fundamental configurations of two solid bodies are shown in Fig. 7.4.

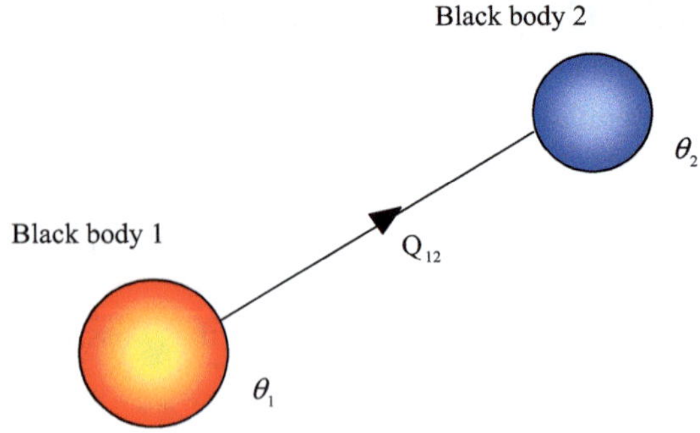

Fig. 7.3 Radiation heat transfer between two black bodies

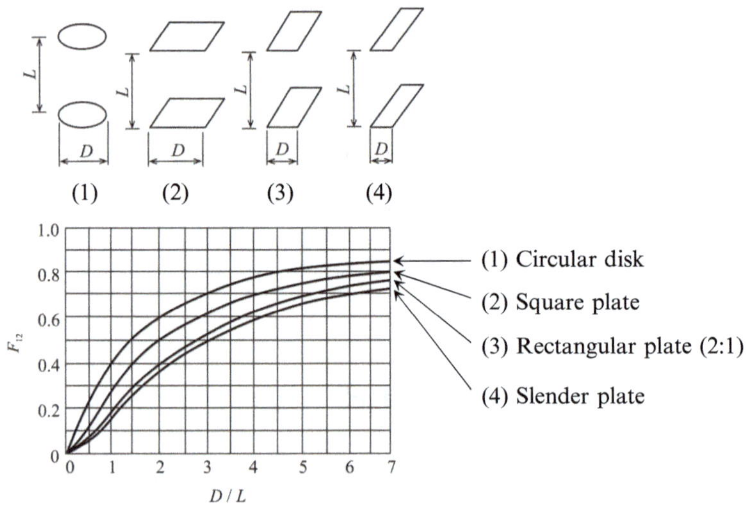

Fig. 7.4 Shape factor, F_{12}, for parallel plates

7.5 Radiation Between Two Surfaces

7.5.1 Shape Factor

The shape factors, F_{12} and F_{21} for $D/L \rightarrow \infty$, are given from Fig. 7.4 as

$$F_{12} = F_{21} = 1 \tag{7.10}$$

The net radiation heat transfer between two very large black flat plates placed in parallel is given by

$$Q_{12} = E_B A_1 = \sigma_B A_1 \left[\left(\frac{\theta_1}{100} \right)^4 - \left(\frac{\theta_2}{100} \right)^4 \right] \tag{7.11}$$

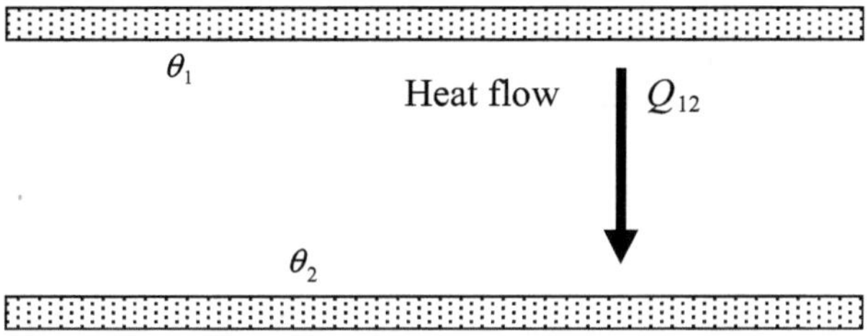

Fig. 7.5 Radiation heat transfer between very large black flat plates placed in parallel

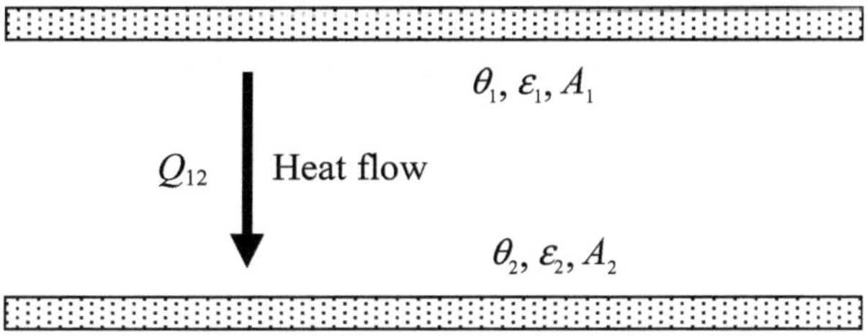

Fig. 7.6 Radiation heat transfer between very large gray flat plates placed in parallel

Exercise 7.3 Calculate the radiation heat flux, $E_B = Q_{12}/A_1$, between two very large black flat plates, 1 and 2, placed in parallel (see Fig. 7.5). The surface temperatures are 1,273 K and 273 K, respectively.

Answer to Exercise 7.3 The emissive power or heat flux can be calculated from Eq. (7.11) to give

$$\begin{aligned} E_B = Q_{12}/A_1 &= \sigma_B \left[\left(\frac{\theta_1}{100} \right)^4 - \left(\frac{\theta_2}{100} \right)^4 \right] \\ &= 5.67 \left[\left(\frac{1,273}{100} \right)^4 - \left(\frac{273}{100} \right)^4 \right] = 5.67(26,261 - 56) \\ &= 1.49 \times 10^5 \ \mathrm{W/m^2} \end{aligned} \tag{7.12}$$

7.5.2 *Radiation Heat Transfer Between Two Gray Flat Plates with No Medium Between Them*

$$Q_{12} = \frac{\sigma_B \left[\left(\frac{\theta_1}{100} \right)^4 - \left(\frac{\theta_2}{100} \right)^4 \right]}{\frac{1-\varepsilon_1}{\varepsilon_1 A_1} + \frac{1}{A_1 F_{12}} + \frac{1-\varepsilon_2}{\varepsilon_2 A_2}} \tag{7.13}$$

where ε_1 is the emissivity of plate 1 and ε_2 is that of plate 2 (see Fig. 7.6).

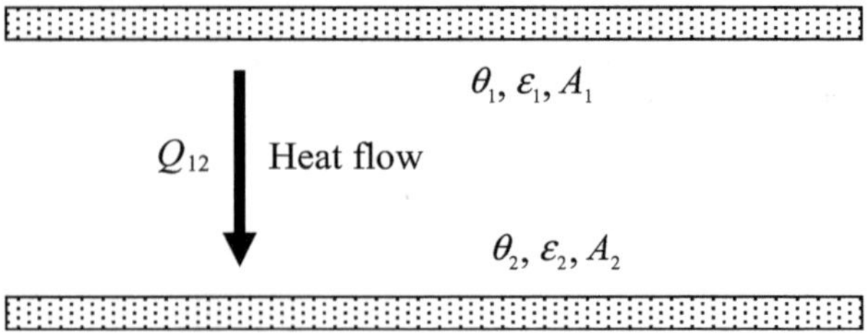

Fig. 7.7 Radiation heat transfer between infinite gray flat plates placed in parallel

7.5.3 Radiation Heat Transfer Between Two Infinite Parallel Planes

When the surface areas, A_1 and A_2, are very large and $A_1 = A_2$, the shape factor is unity ($F_{12} = 1$). Under this condition Eq. (7.13) reduces to

$$Q_{12} = \frac{\sigma_B A_1 \left[\left(\frac{\theta_1}{100} \right)^4 - \left(\frac{\theta_2}{100} \right)^4 \right]}{\frac{1}{\varepsilon_1} + \frac{1}{\varepsilon_2} - 1} \tag{7.14}$$

Exercise 7.4 Two infinite planes, 1 and 2, are placed in parallel, as shown in Fig. 7.7. Calculate the heat transferred by radiation per unit time and unit area, Q_{12} (W/m^2). The temperatures on planes 1 and 2 are 600 K and 300 K, respectively. Their emissivity values are equal, $\varepsilon_1 = \varepsilon_2 = 0.20$.

Answer to Exercise 7.4 The radiative heat transfer, Q_{12}, can be calculated from Eq. (7.14):

$$\begin{aligned} Q_{12} &= \frac{\sigma_B A_1 \left[\left(\frac{\theta_1}{100} \right)^4 - \left(\frac{\theta_2}{100} \right)^4 \right]}{\frac{1}{\varepsilon_1} + \frac{1}{\varepsilon_2} - 1} = \frac{5.67 \times 1 \times \left[\left(\frac{600}{100} \right)^4 - \left(\frac{300}{100} \right)^4 \right]}{\frac{1}{0.20} + \frac{1}{0.20} - 1} \\ &= \frac{5.67 \times 1 \times (1{,}296 - 81)}{5 + 5 - 1} = 765\,\mathrm{W/m^2} \end{aligned} \tag{7.15}$$

Exercise 7.5 In Exercise 7.4 a third infinite plate is placed between the two infinite planes, 1 and 2. Plate 3 is also placed in parallel to the remaining plates, as shown in Fig. 7.8. Calculate the heat transferred by radiation per unit time and unit area, Q_{12} (W/m^2). The temperatures on planes 1 and 2 are $\theta_1 = 600$ K and θ_2=300 K, respectively. The emissivity values of the plates are equal; thus, $\varepsilon_1 = \varepsilon_2 = \varepsilon_3 = 0.20$.

Fig. 7.8 Shield effect on radiation heat transfer between infinite gray flat plates placed in parallel

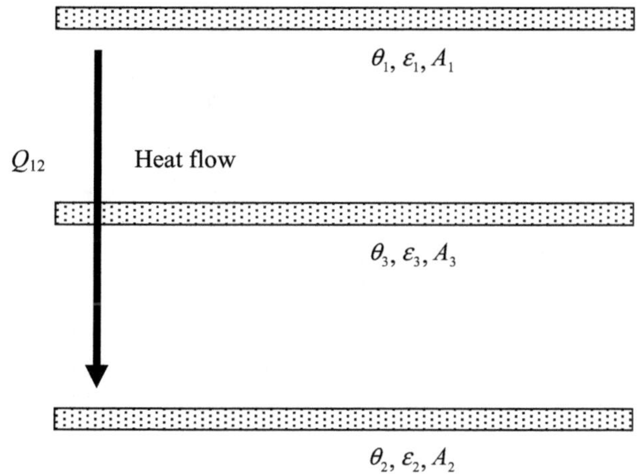

Answer to Exercise 7.5 The temperature of plate 3 is not known and denoted by θ_3. The radiative heat transfer, Q_{12}, can be calculated from Eq. (7.14):

$$Q_{12} = Q_{13} = Q_{32} = \frac{\sigma_B A_1 \left[\left(\frac{\theta_1}{100} \right)^4 - \left(\frac{\theta_3}{100} \right)^4 \right]}{\frac{1}{\varepsilon_1} + \frac{1}{\varepsilon_3} - 1}$$

$$= \frac{\sigma_B A_1 \left[\left(\frac{\theta_3}{100} \right)^4 - \left(\frac{\theta_2}{100} \right)^4 \right]}{\frac{1}{\varepsilon_3} + \frac{1}{\varepsilon_2} - 1} \tag{7.16}$$

The relationship between θ_1, θ_2, and θ_3 is expressed by

$$\left(\frac{\theta_1}{100} \right)^4 - \left(\frac{\theta_3}{100} \right)^4 = \left(\frac{\theta_3}{100} \right)^4 - \left(\frac{\theta_2}{100} \right)^4 \tag{7.17}$$

The temperature, θ_3, becomes

$$\left(\frac{\theta_3}{100} \right)^4 = \frac{1}{2} \left[\left(\frac{\theta_1}{100} \right)^4 + \left(\frac{\theta_2}{100} \right)^4 \right] \tag{7.18}$$

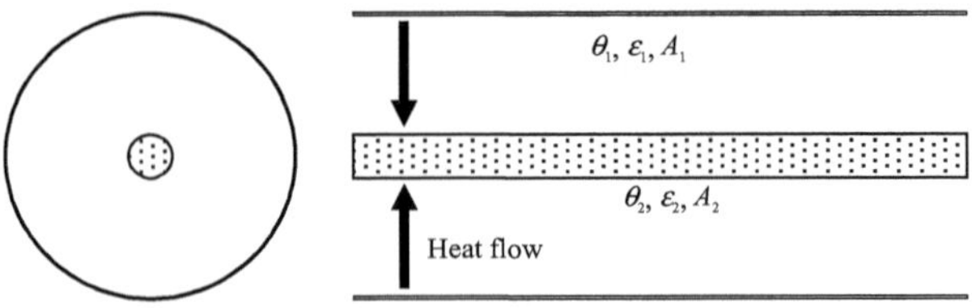

Fig. 7.9 Radiation heat transfer between very long coaxial gray cylinders

Accordingly, Q_{12} is given by

$$Q_{12} = \frac{\frac{1}{2}\sigma_B A_1\left[\left(\frac{\theta_1}{100}\right)^4 - \left(\frac{\theta_2}{100}\right)^4\right]}{\frac{1}{\varepsilon_1} + \frac{1}{\varepsilon_3} - 1}$$

$$= \frac{\frac{1}{2} \times 5.67 \times 1 \times (1{,}296 - 81)}{5 + 5 - 1} = 383\ \mathrm{W/m^2} \tag{7.19}$$

It should be noted that the heat flux is one-half of that derived in the previous exercise. That is, inserting a plate between two heated plates can shield radiation heat transfer. When the number of shields is n, the following shield effect can be obtained:

$$Q_{12}(\text{with } n \text{ shields}) = \frac{1}{n+1} Q_{12}(\text{without shields}) \tag{7.20}$$

7.5.4 Radiation Heat Transfer Between Two Concentric Long Coaxial Cylinders

The configuration considered is shown in Fig. 7.9. The heat transferred by radiation is given by

$$Q_{12} = \frac{\sigma_B A_1\left[\left(\frac{\theta_1}{100}\right)^4 - \left(\frac{\theta_2}{100}\right)^4\right]}{\frac{1}{\varepsilon_1} + \frac{A_1}{A_2}\left(\frac{1}{\varepsilon_2} - 1\right)} \tag{7.21}$$

When A_2 is much larger than A_1, Eq. (7.21) becomes

$$Q_{12} = \varepsilon_1 \sigma_B A_1\left[\left(\frac{\theta_1}{100}\right)^4 - \left(\frac{\theta_2}{100}\right)^4\right] \tag{7.22}$$

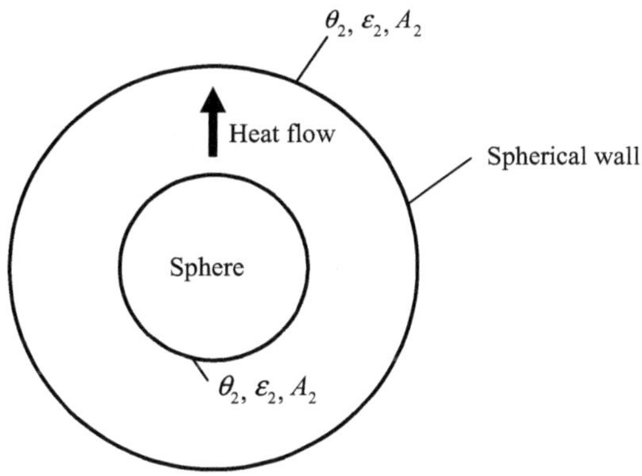

Fig. 7.10 Radiation heat transfer between concentric two spherical surfaces

7.5.5 *Radiation Heat Transfer Between Two Concentric Spherical Surfaces*

Consider the configuration in Fig. 7.10. Equation (7.21) is still valid in this case; thus,

$$Q_{12} = \frac{\sigma_B A_1 \left[\left(\frac{\theta_1}{100} \right)^4 - \left(\frac{\theta_2}{100} \right)^4 \right]}{\frac{1}{\varepsilon_1} + \frac{A_1}{A_2} \left(\frac{1}{\varepsilon_2} - 1 \right)} \tag{7.23}$$

7.6 Relevant Dimensionless Parameters in Heat Transfer

7.6.1 *Grashof Number*

$$Gr = \frac{g\rho^2 \beta \Delta\theta L^3}{\mu^2} \quad \left(\frac{\text{Inertial force}}{\text{Viscous force}} \times \frac{\text{Buoyancy force}}{\text{Viscous force}} \right) \tag{7.24}$$

where g is the acceleration due to gravity (m/s^2), ρ is the density of fluid (kg/m^3), β is the coefficient of thermal volume expansion (1/K), $\Delta\theta$ is the temperature difference (K), L is the representative length (m), and μ is the dynamic viscosity of fluid (kg/m s). This dimensionless number is relevant to natural convection.

7.6.2 Nusselt Number

$$Nu = \frac{\alpha L}{\lambda} \left(\frac{\text{Convection heat transfer}}{\text{Conduction heat transfer}}\right) \tag{7.25}$$

where α is the heat-transfer coefficient (W/m^2 K), L is the representative length (m), and λ is the thermal conductivity (W/m K). The Nusselt number is typically used in forced convection.

7.6.3 Peclet Number

$$Pe = \frac{LV}{\kappa} \left(\frac{\text{Heat transfer by bulk motion}}{\text{Convection heat transfer}}\right) \tag{7.26}$$

where L is the representative length (m), V is the representative velocity of fluid (m/s), and κ is the thermal diffusivity (m^2/s). Peclet number is also used in heat convection.

7.6.4 Prandtl Number

$$Pr = \frac{\nu_f}{\kappa} \left(\frac{\text{Momentum diffusivity}}{\text{Thermal diffusivity}}\right) \tag{7.27}$$

where ν_f is the kinematic viscosity of fluid (m^2/s) and κ is the thermal diffusivity (m^2/s). The Prandtl number is relevant in heat convection.

7.6.5 Rayleigh Number

$$Ra = Gr\,Pr(\text{Grashof number} \times \text{Prandtl number}) \tag{7.28}$$

where Gr is the Grashof number (–) and Pr is the Prandtl number (–). Rayleigh number is typically used to describe laminar–turbulent transition in natural convection.

Part III
Mass Transfer

Chapter 8
Diffusion and Mass Transfer

Abstract Diffusion is the transport of material due to concentration gradients or more precisely due to gradients in the chemical potential. Mass transfer is commonly described as diffusional phenomena in the presence of convective motion. Diffusion and mass transfer play a significant role in many materials processing operations. In metal extraction and refining, the chemical changes are usually associated with the transport of the reactants to the reaction sites and the transport of products away from the reaction site. The overall rates of the processes are often mass transfer-controlled, because the chemical kinetics tend to be fast at elevated temperatures. Diffusion and mass transfer are also of crucial importance in affecting the structure of solid products. Typical examples include segregation phenomena in castings and impurity distribution in crystal growth. This chapter introduces the mathematical formulation of diffusion in homogeneous and porous media and discusses associated phenomena including mass diffusion and chemical reaction. The boundary conditions required to solve mass transfer problems are presented. The chapter concludes with diagnostic examples on the basic principles presented.

Keywords: Diffusion • Fick's law • Mass diffusion • Mass flux • Mass transfer

8.1 Introduction

Diffusion is the transport of material due to concentration gradients or more precisely due to gradients in the chemical potential. Mass transfer is commonly described as diffusional phenomena in the presence of convective motion.

Diffusion and mass transfer play a significant role in many materials processing operations. In metal extraction and refining, the chemical changes are usually associated with the transport of the reactants to the reaction sites and the transport of products away from the reaction site. The overall rates of the processes are often mass transfer-controlled, because the chemical kinetics tend to be fast at elevated temperatures. Diffusion and mass transfer are also of crucial importance in

M. Iguchi and O.J. Ilegbusi, *Basic Transport Phenomena in Materials Engineering*,
DOI 10.1007/978-4-431-54020-5_8,

affecting the structure of solid products. Typical examples include segregation phenomena in castings and impurity distribution in crystal growth. This chapter presents basic principles of diffusion and mass transfer including processes in homogeneous and porous media and mass transfer involving chemical reaction, as well formulating the boundary conditions.

8.2 Formulating Diffusion in Homogeneous Media

For a homogeneous medium such as a gas or a liquid, Fick's law of diffusion is given as

$$J_{\mathrm{A}} = D_{\mathrm{AB}} \nabla C_{\mathrm{A}} \tag{8.1}$$

where J_{A} is the molar flux of species A relative to the molar average velocity, C_{A} is the concentration of the diffusing species, D_{AB} is the binary diffusion coefficient.

This equation can be expressed in terms of N_{A}, the flux for a fixed coordinate system to yield

$$N_{\mathrm{A}} = x_{\mathrm{A}}(N_{\mathrm{A}} + N_{\mathrm{B}}) - D_{\mathrm{AB}} \nabla C_{\mathrm{A}} \tag{8.2}$$

where x_A is the mole fraction of species A.

The equation of continuity for a given species, written in terms of these fluxes, takes the following form:

$$\frac{\partial C_{\mathrm{A}}}{\partial t} + \nabla \cdot N_{\mathrm{A}} = R_{\mathrm{A}} \tag{8.3}$$

where R_{A} is the rate of production of species A by homogeneous chemical reaction.

Upon combining Eq. (8.3) with the flux relationships in the absence of bulk motion for which $N_{\mathrm{A}} = -N_{\mathrm{B}}$, or when $x_{\mathrm{A}} \to 0$, the following equation is obtained, assuming constant density and constant diffusion coefficient [1]:

$$\frac{\partial C_{\mathrm{A}}}{\partial t} = D_{\mathrm{A}} \nabla^2 C_{\mathrm{A}} + R_{\mathrm{A}} \tag{8.4}$$

It should be noted that the form of Eq. (8.4) is very similar to that given for the equivalent heat conduction problems in the previous Part II, with the reaction term replacing the heat generation term.

There are a number of complicating factors that have to be considered in the modeling of diffusion phenomena. These factors include the following:

1. For gas phase systems, in the absence of equimolar counter-diffusion and when the transferred species is at high mole fraction levels, the equations do get modified as a result of bulk flow due to diffusion.
2. In multicomponent systems, there can be a coupling between the fluxes of the individual species. Thus, the relationship between the fluxes and the gradients takes the following form:

$$\nabla x_i = \sum_{j=1}^{n} \frac{1}{D_{ij}} \left(x_i N_j - x_j N_i \right) \tag{8.5}$$

where N_i, N_j are the fluxes and D_{ij} are the binary diffusivities.

This set of equations is, of course, quite complex, but may be required for the faithful representation of systems involving several species with a disparate molecular weight.

Yet other complications may arise within the same framework. When diffusion processes take place in a steep temperature gradient, thermal diffusion or the Soret effect may have to be considered also. Under these conditions, the diffusive flux may be written as

$$j_{\mathrm{A}}^{T} = -\frac{c^2}{\rho} M_{\mathrm{A}} M_{\mathrm{B}} D_{\mathrm{AB}} \frac{k_T}{T} \nabla T \tag{8.6}$$

where k_T is the Soret coefficient, M_{A} and M_{B} are the molecular weights of species A and B, respectively, and c is the concentration of the diffusing species.

Such problems may arise, for example, in levitation melting and could be a factor in chemical vapor deposition systems.

For a one-dimensional problem, Eq. (8.5) reduces to

$$\frac{dx_{\mathrm{A}}}{dz} = -\frac{k_T}{T} \frac{dT}{dz} \tag{8.7}$$

Note that R_{A} in Eq. (8.3), the rate of homogeneous chemical reaction in the system, cannot be defined independently of the other process parameters. Indeed, in general R_{A} will depend on the concentration of all the reactants and also on the temperature and pressure. It follows that additional expressions have to be invoked in order to specify R_{A}.

8.3 Diffusion in Porous Media

In many practical applications, it is necessary to deal with diffusion in porous media, i.e., solids with pores or interconnected voids. Examples of such situations include the oxidation of metals, the reduction of metal oxides, the drying of ceramics, and the combustion of porous solid fuels.

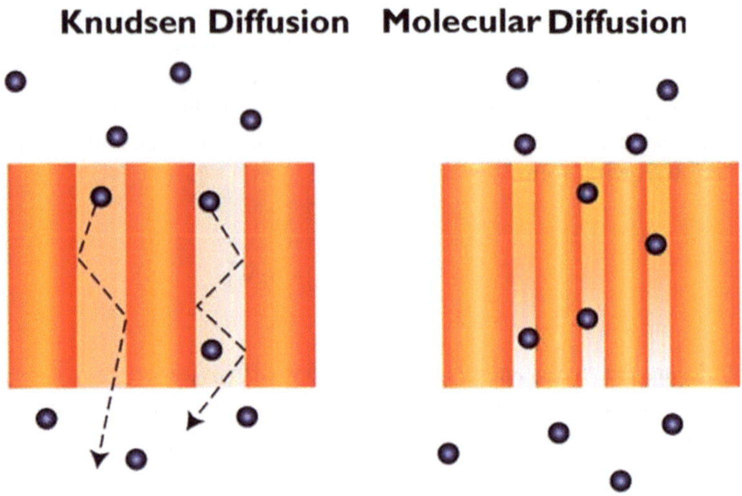

Fig. 8.1 Diffusion in porous medium [2]

Figure 8.1 shows a schematic sketch of diffusion within porous media. The presence of the pores introduces the following complications:

- Not all the volume is available for diffusion.
- The diffusion path will become tortuous, i.e., non-straight.
- When the mean free path of the molecules is of the same order of magnitude as the size of the pores, Knudsen diffusion, rather than molecular diffusion, will become the dominant mechanism [3]. Generally this will occur when the pore diameter approaches about 1 μm.

The equations developed for bulk diffusion may be used for diffusion in porous media, but the binary diffusion coefficient, D_{AB}, has to be replaced by $D_{A,eff}$, the effective diffusion coefficient. These two quantities are related by the following expression:

$$\frac{1}{D_{A,eff}} = \frac{1}{D_{AK}} + \frac{1}{D_{A,B,eff}} \tag{8.8}$$

$$D_{A,B,eff} = \frac{\varepsilon}{\tau} D_{A,B} \tag{8.9}$$

where ε is the porosity, τ is the tortuosity, D_{AK} is the Knudsen diffusion coefficient.

The tortuosity factor essentially corrects for the path being generally longer than a straight line. For example, it is typically in the range 1.5–5 for porous inert solids. A good discussion of some of these issues is available in several specialized texts [3–5].

The governing equations for convective mass transfer take the following form:

$$\frac{DC_{\mathrm{A}}}{Dt} = \nabla D_{\mathrm{A-B}} \nabla C_{\mathrm{A}} + R_{\mathrm{A}} \tag{8.10}$$

For turbulent flow, identical relationships will apply, but here the diffusivity has to be replaced by its effective value, $D_{\mathrm{AB}(l,t)}$, which combines the molecular and turbulent contributions; thus,

$$\frac{DC_{\mathrm{A}}}{Dt} = \nabla D_{\mathrm{A-B}(l,t)} \nabla C_{\mathrm{A}} + R_{\mathrm{A}} \tag{8.11}$$

By analogy with convective heat transfer, both forced and natural convection may be considered. In forced convection, the flow field is externally imposed, e.g., through the use of mixers, gas bubbles, and electromagnetic force fields. In the case of natural convection, the flow is driven by the density gradients, which in turn are caused by the concentration gradients.

In the consideration of mass transfer, an additional complication may arise, namely that velocity fields produced by thermal effects will produce convective mass transfer. Indeed, such problems are of considerable importance in crystal growth.

The Peclet number for mass transfer may also be defined in analogy with convective heat transfer as

$$Pe = \frac{UL}{D_{\mathrm{A-B}}} \tag{8.12}$$

where U is the characteristic velocity, L is the characteristic length, $D_{\mathrm{A-B}}$ is the diffusion coefficient.

In liquid systems, $D_{\mathrm{A-B}}$ is generally quite small, on the order of 10^{-9} to 10^{-8} m^2/s. This has two very important consequences:

- In most cases, high levels of turbulence, i.e., turbulent diffusivities of the order of 10^{-2} m^2/s, are required to bring about the rapid homogenization of the system. It is very difficult to homogenize a vessel under laminar flow conditions.
- The very low numerical value of the diffusivity implies that even quite small fluid velocities, say, of the order of 1 mm/s or less, may have a very marked effect in modifying the concentration profiles. Indeed, even for such low values of the characteristic velocity, convection effects may dominate.

8.4 Mass Transfer with Chemical Reaction

Mass transfer in a chemically reacting system will be illustrated with the processing of metal oxide thin film through surface hydrolysis of metal chloride on a heated substrate as illustrated in Fig. 8.2 [6]:

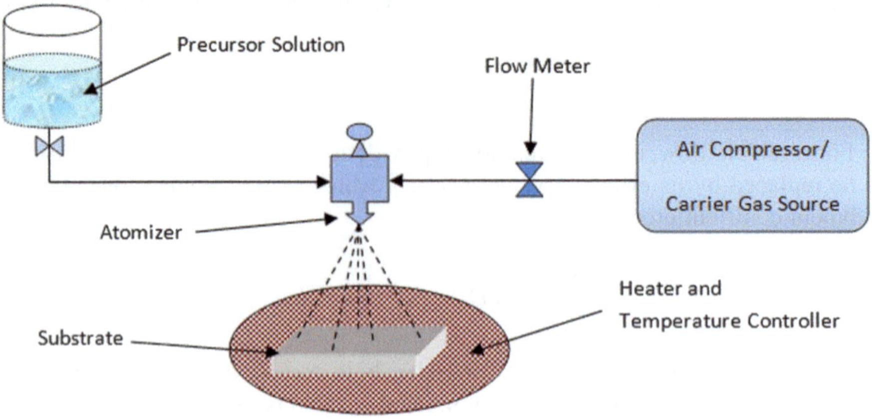

Fig. 8.2 Schematic of chemical pyrolysis system involving surface hydrolysis of metal chloride on a heated substrate [6]

The process follows the reaction [7]

$$XCl_m + nH_2O \rightarrow XO_n + mHCl \tag{8.13}$$

where X is the metal such as Sn, Zn, and Cu of the oxide films.

8.4.1 Mass Diffusion

The local mass fraction of each species Y_i can be predicted through the solution of a convention–diffusion equation for the ith species. The conservation equation takes the following general form:

$$\frac{\partial}{\partial t}(\rho Y_i) + \nabla \cdot (\rho \vec{v} Y_i) = -\nabla \cdot \vec{J_i} + R_i + S_i \tag{8.14}$$

where R_i is the net rate of production of species i by chemical reaction and S_i is the rate of reaction n by addition from the dispersed phase. The parameter J_i is the diffusion flux of species i, which arises due to gradients of concentration and temperature. Fick's law is used to model mass diffusion due to concentration gradients, and the flux can be written as

$$\vec{J_i} = -\rho D_{i,m} \nabla Y_i - D_{T,i} \frac{\nabla T}{T} \tag{8.15}$$

Here $D_{i,m}$ is the mass diffusion coefficient for species i in the mixture, and $D_{T,i}$ is the thermal diffusion coefficient.

8.4.2 Chemical Reaction Model

The deposition rate G follows the Arrhenius law:

$$G = A\exp(-E_a/RT) \tag{8.16}$$

where A is a constant, E_a is the apparent activation energy, R is the gas constant, and T is the deposition temperature. The net source of chemical species i due to reactions is computed as the sum of the Arrhenius reaction sources over the NR reactions that the species participate in:

$$R_i = M_{w,i}\sum_{r=1}^{N_R}\widehat{R}_{i,r} \tag{8.17}$$

where $M_{w,i}$ is the molecular weight of species i and $\widehat{R}_{i,r}$ is the Arrhenius molar rate of creation/destruction of species i in reaction r. The reaction occurs in the continuous phase at the wall surface.

Consider the rth reaction written in general form as follows:

$$\sum_{i=1}^{N} v'_{i,r}\mathcal{M}_i \underset{k_{\mathrm{b,r}}}{\overset{k_{\mathrm{f,r}}}{\rightleftharpoons}} \sum_{i=1}^{N} v''_{i,r}\mathcal{M}_i \tag{8.18}$$

where N is the Number of chemical species in the system, $v'_{i,r}$ is the stoichiometric coefficient for reactant i in reaction r, $v''_{i,r}$ is the stoichiometric coefficient for product i in reaction r, $\mathcal{M}_i$ is the symbol denoting species i, $k_{\mathrm{f,r}}$ is the forward rate constant for reaction r, $k_{\mathrm{b,r}}$ is the backward rate constant for reaction r.

This equation is valid for both reversible and nonreversible reactions. For nonreversible reactions, the backward rate constant, $k_{\mathrm{b,r}}$, is omitted.

Based on Eq. (8.16), the forward rate constant for each $k_{\mathrm{f,r}}$ is computed using the Arrhenius expression:

$$k_{\mathrm{f,r}} = A_r T^{\beta_{\mathrm{re}} - E_{\mathrm{r}}/RT} \tag{8.19}$$

where A_r is the pre-exponential factor, β_r is the temperature exponent, E_r is the activation energy for the reaction (J/kmol), R is the universal gas constant (J/kmol).

Since this is an endothermic reaction, a hot substrate provides the thermal energy for the thermal decomposition and subsequent recombination of the constituent species.

8.5 Boundary Conditions for Mass Transfer

The boundary conditions required to describe mass transfer problems closely parallel those of heat transfer, although there are some differences that will be discussed subsequently. These conditions are:

a. Fixed concentration at the bounding surfaces

$$C = C_{\mathrm{o}};\ y = y_{\mathrm{o}} \tag{8.20}$$

or

$$C = C(x, y, z) \text{ at } f(x, y, z) \tag{8.21}$$

b. Mass flux at the bounding surface

$$-D_{\mathrm{A-B}}\frac{\partial C_{\mathrm{A}}}{\partial y} = n(t) \tag{8.22}$$

or, more generally at the surface Γ,

$$-D_{\mathrm{A-B}}\frac{\partial C_{\mathrm{A}}}{\partial n} = \mathbf{n}(t);\quad \Gamma = \Gamma(x, y, z) \tag{8.23}$$

c. Specify a relationship between the flux and the concentration at the bounding surface

$$-D_{\mathrm{A-B}}\frac{\partial C_{\mathrm{A}}}{\partial y} = f(C_{\mathrm{A}});\ y = y_0 \tag{8.24}$$

or

$$-D_{\mathrm{A-B}}\frac{\partial C_{\mathrm{A}}}{\partial n} = f(C_{\mathrm{A}});\ \Gamma = \Gamma(x, y, z) \tag{8.25}$$

A particular example of the latter may include

$$f(C_{\mathrm{A}}) = h_{\mathrm{D}}\left(C_{\mathrm{A}} - C_{\mathrm{A},0}\right) \tag{8.26}$$

expressing a convective boundary condition in which h_{D} is the mass transfer coefficient.

A special application of this concept in materials processing operation is the mass transfer from some interface into a well-mixed solution. The amount transferred can be assumed to be proportional to the concentration difference and the interfacial area thus:

Rate of mass transferred = k (Interfacial area) (Concentration difference)

where k is a mass transfer coefficient. By dividing both sides of this equation by the area, the above relationship can be written as

$$N = k[c_{1i} - c_1] \tag{8.27}$$

where N is the flux at the interface and c_{1i} and c_1 are the concentrations at the interface and in the bulk solution, respectively [8].

Alternatively, the boundary condition may express the fact that the rate of diffusion to the interface must equal the rate of chemical reaction. In general, the reaction term will be a nonlinear function of the concentration; thus,

$$-D_{\mathrm{A-B}} \frac{\partial C_{\mathrm{A}}}{\partial n} = R_{\mathrm{A}}(C_{\mathrm{A}}) \tag{8.28}$$

The heat transfer and mass transfer characteristics are generally quite different in multiphase systems.

d. At a gas–solid interface in which the gas (species A) dissolves in the solid (species B) and a homogenous solution is formed, gas transport can be assumed to be a diffusion process being independent of the solid structure. If the solid could be considered a uniform substance, the concentration in the solid at the interface may be obtained with the use of the property *solubility, S.* The expression relating the partial pressure of gas right next to the interface P_{A}, the molar concentration of the gas in the solid interface $C_{\mathrm{A}}(0)$, and solubility S is [9]

$$C_{\mathrm{A}}(0) = S \times P_{\mathrm{A}} \tag{8.29}$$

The unit of the solubility, S, is kilomoles of species A per cubic meter of solid per bar partial pressure of A.

e. At the interface of two fluids, we express the existence of equilibrium at the interface:

$$\frac{C_{\mathrm{A/I}}}{C_{\mathrm{A/II}}} = K_{\mathrm{E}} \tag{8.30}$$

where K_E is the equilibrium partition coefficient. In general this condition entails a discontinuity in the concentration versus distance curves.

At the same interface ($y = y_0$), the continuity of the fluxes implies

$$-D_{A-B,I}\left(\frac{\partial C_A}{\partial y}\right)_I = -D_{A-B,II}\left(\frac{\partial C_A}{\partial y}\right)_{II} \tag{8.31}$$

f. On the analogy of heat transfer for phase change or moving boundary problems, the boundary condition for mass transfer, e.g., for the growth of gas bubbles in liquids, may be expressed as

$$\rho_M \frac{dY}{dt} = D_{A-B}\frac{\partial C_A}{\partial y}, \quad y = Y \tag{8.32}$$

This equation expresses the fact that the rate of diffusive transfer to the interface must equal the molar rate of phase growth.

Exercise 8.1 Consider water evaporating into initially dry air in a closed vessel. The vessel is isothermal at 25 °C, such that the water vapor pressure is 23.8 mm Hg. The vessel has a total volume of 19.2 L and cross-sectional area of 100 cm^2 and contains 0.8 L of water. If the mass transfer rate is 3.3×10^{-8} mol/cm^2 s, how long will it take for the air to become 5 % saturated?

Answer to Exercise 8.1 At time t the flux at can be calculated thus:

$$N_1 = \frac{(\text{Vapor concentaration})(\text{Air volume})}{(\text{Liquid area})(\text{Time})}$$

$$= \frac{0.05\left(\frac{23.8}{760}\right)\left(\frac{1\,\text{mol}}{22.4\,\text{L}}\right)\left(\frac{273}{298}\right)(18.4\,\text{L})}{(100\,\text{cm}^2)(t)} \times 3.3 \times \frac{10^{-8}\text{mol}}{\text{cm}^2 s} \quad t = 360\,\text{s} = 6\,\text{min}$$

Exercise 8.2 An oxygen bubble initially 0.1 cm diameter is injected into a bath of continuously agitated water. If the mass transfer coefficient is 1.76×10^{-3} cm/s, estimate how long it will take for the bubble diameter to reduce to 0.064 cm.

Answer to Exercise 8.2 Applying mass balance to the bubble rather than the surrounding solution gives

$$\frac{d}{dt}\left(c_1\frac{4}{3}\pi r^3\right)\text{AN}_1 = -4\pi r^2 k[c_1(\text{sat}) - 0]$$

where c_1 is the oxygen concentration in the bubble (= 1 mol/22.4 L at standard conditions), $c_1(\text{sat})$ is the oxygen concentration at saturation in water (about 1.5×10^{-3}

moles per liter under similar conditions), A is the surface area of the bubble, N is the mass flux, and k is mass transfer coefficient. Thus,

$$\frac{dr}{dt} = -k\frac{c_1(\text{sat})}{c_1} = -0.034k$$

Integrating subject to the initial condition

$$t = 0, \ r = 0.05\,\text{cm}$$

gives

$$r = 0.05\,\text{cm} - 0.034\ kt$$

Inserting the numerical values given,

$$0.032\,\text{cm} = 0.05\,\text{cm} - 0.034\left(1.76 \times 10^{-3}\,\text{cm/s}\right)(t) -> t = 300\ \text{s} = 5\,\text{min}$$

Exercise 8.3 A mixture of He and N_2 gas is contained in a pipe at 298 K and 1 atm total pressure which is constant throughout. At one end of the pipe at point (1) the partial pressure p_{A1} of He is 0.60 atm and at the other end 0.2 m, $p_{A2} = 0.20$ atm. Calculate the flux of He at steady state if D_{AB} of the He–N_2 mixture is 0.687 cm^2/s. (Hint: Use the perfect-gas law to find a relationship between concentration and partial pressure).

Answer to Exercise 8.3

$$PV = nRT$$

$$\frac{n}{V} = \frac{P}{RT} = c$$

The flux (J^*_{AZ}) is calculated thus:

$$J^*_{AZ} = -D_{AB}\frac{dc_A}{dz}$$

$$J^*_{AZ}\int_{z_1}^{z_2} dz = -D_{AB}\int_{c_{A1}}^{c_{A2}} dc_A$$

$$J^*_{AZ} = \frac{D_{AB}(c_{A2} - c_{A1})}{z_2 - z_1}$$

$$c_{A1} = \frac{P_{A1}}{RT} = \frac{n_A}{V}$$

$$J^*_{AZ} = \frac{D_{AB}(P_{A2} - P_{A1})}{RT(z_2 - z_1)}$$

$$P_{A1} = 6.08 \times 10^4 Pa$$

$$P_{A2} = 2.027 \times 10^4 Pa$$

$$J^*_{AZ} = \frac{D_{AB}(P_{A2} - P_{A1})}{RT(z_2 - z_1)} = \frac{(0.687 \times 10^{-4})(6.08 \times 10^4 - 2.027 \times 10^4)}{(8314)(298)(0.2 - 0)}$$

$$J^*_{AZ} = 5.63 \times 10^{-6} \frac{\text{kg mol A}}{\text{s m}^2}$$

Exercise 8.4 Oxygen gas is maintained at pressures of 2 bars and 1 bar on opposite sides of a rubber membrane that is 2.5 mm thick, and the entire system is at 25 °C. What is the molar diffusive flux of oxygen through the membranes? Assume for oxygen at 298 K: $D_{AB} = 0.21 \times 10^{-9}$ m^2/s and solubility in rubber $S = 3.12 \times 10^{-3}$ k mol/m^3 bar. Repeat the calculation for rubber membrane thicknesses of 1.5, 3.5, and 4.5 and plot the molar diffusive flux versus thickness of the membrane. Comment on the result.

Answer to Exercise 8.4

$$N''_{A,x} = J^*_{A,x} = -D_{AB}\frac{dC_A}{dx}$$

$$C_A(0) = Sp_{A,1} = 6.24 \times 10^{-3}\frac{\text{k mol}}{\text{m}^3}$$

$$C_A(L) = Sp_{A,2} = 3.12 \times 10^{-3}\frac{\text{k mol}}{\text{m}^3}$$

Hence,

$$N''_{A,x} = 0.21 \times \frac{10^{-9}\text{m}^2}{S}\frac{\left(\frac{(6.24 \times 10^{-3} - 3.12 \times 10^{-3})\text{k mol}}{\text{m}^3}\right)}{0.0025\,\text{m}}$$

$$N''_{A,x} = 0.21 \times \frac{10^{-9}\text{m}^2}{\text{s}}\frac{\left(\frac{(6.24 \times 10^{-3} - 3.12 \times 10^{-3})\text{k mol}}{\text{m}^3}\right)}{0.0025\,\text{m}}$$

$$N''_{A,x} = 2.62 \times 10^{-10}$$

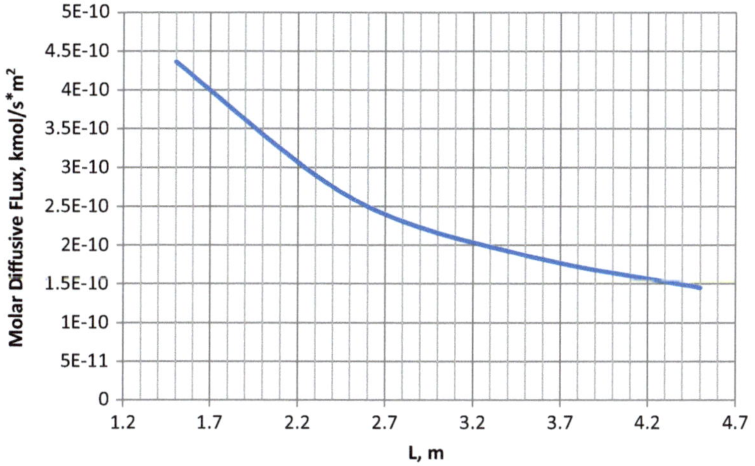

Fig. 8.3 Variation of diffusive flux with membrane thickness

Figure 8.3 shows the calculated flux variation with rubber thickness. The results show that as the thickness of the rubber membrane increases, the molar diffusive flux decreases, in the consensus of Fick's law.

References

1. Bird RB, Stewart WE, Lightfoot EN (1960) Transport phenomena. Wiley, New York
2. The Cooperative Research Centre for Greenhouse Gas Technologies CO_2 capture/separation technologies. http://www.co2crc.com.au/aboutccs/cap_membranes.html
3. Geankoplis CJ (1993) Transport process and unit operations. Prentice-Hall, Englewood Cliffs
4. Szekely J, Themelis NJ (1971) Rate phenomena in process metallurgy. Wiley, New York
5. Poirier DR, Geiger GH (1994) Transport phenomena in materials processing. The Minerals. Metals and Materials Society (TMS), Warrendale
6. Khatami N, Ilegbusi O (2011) Synthesis and modeling of mixed-oxide nano-composite sensor film. In: Proceedings of the ASME 2011 international mechanical engineering congress & exposition IMECE2011, Denver, November 2011
7. Hartnagel HL (1995) Semiconducting transparent thin films. Institute of Physics Publishing, Bristol and Philadelphia
8. Cussler EL (2009) Diffusion: mass transfer in fluid systems. Cambridge University Press, Cambridge, New York
9. Incropera FP, DeWitt DP (1996) Fundamentals of heat and mass transfer, 4th edn. Wiley, INC

Part IV
Multiphase Flow

Chapter 9
Gas–Liquid Two-Phase Flow

Abstract Bubble and liquid flow characteristics of bubble-driven gas–liquid two-phase flows observed in various types of reactors in materials processing operations are introduced with many exercises. Heat and mass transfer from a solid body immersed in a molten metal bath is also discussed.

Keywords Bubble frequency • Bubble Reynolds number • Bubble rising velocity • Electroresistivity probe • Gas holdup • Half-value radius • Two-fluid model • Void fraction

9.1 Introduction

Gas injection techniques have widely been used in the steelmaking processes [1–15], as shown in Fig. 9.1. The objectives of gas injection can be classified into the following five categories:

(a) Bath agitation due to buoyancy forces acting on bubbles generated in a bath. The so-called gas-lift effect plays a key role for inducing liquid motion.
(b) Removal of impurities in molten steel including carbon, sulfur, and phosphor.
(c) Removal of nonmetallic inclusions such as alumina (Al_2O_3) mainly in the tundish.
(d) Suppression of alumina attachment to the inner wall of the immersion nozzle in a continuous casting process. Argon is typically introduced into the immersion nozzle through a porous plug attached near the sliding gate.
(e) Enhancement of scrap melting.

The shape, size, and number of bubbles and their dispersion patterns in molten steel largely determine the refining efficiency and the removal efficiency of non-metallic inclusions. A thorough understanding of these characteristics is essential for improvement of conventional processes and the development of novel agitation methods. This chapter will introduce the characteristics of gas–liquid two-phase

M. Iguchi and O.J. Ilegbusi, *Basic Transport Phenomena in Materials Engineering*,
DOI 10.1007/978-4-431-54020-5_9,

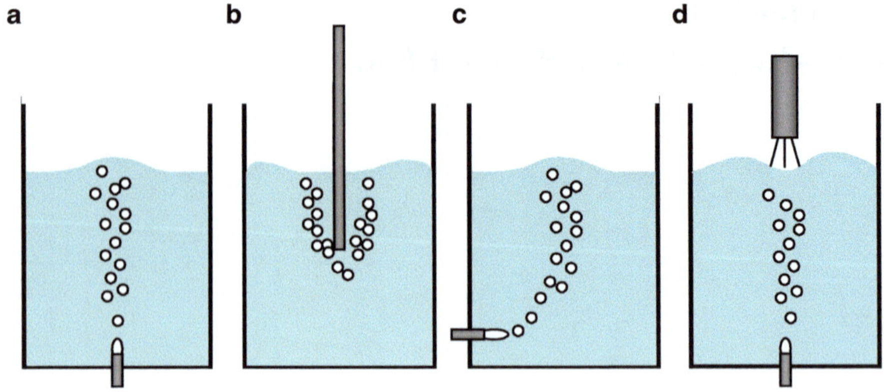

Fig. 9.1 Gas injection types. (**a**) Bottom blowing, G. (**b**) Top blowing. (**c**) Side blowing. (**d**) Top and bottom blowing

flows relevant to real steelmaking processes and water model experiments. Water is commonly used as a model liquid for molten steel because their kinematic viscosities are nearly equal. In addition, water is transparent and, hence, useful for flow visualization.

9.2 Fluid Mechanics

9.2.1 Basic Parameters

Gas–liquid two-phase flows in pipes and ducts have been mostly investigated in mechanical engineering and atomic power engineering, while similar flows in vessels and reactors occur in materials and chemical engineering. The technical terms used in these engineering fields are not necessarily standardized. Important parameters describing the behaviors of gas–liquid two-phase flows are introduced in this section [8].

9.2.1.1 Void Fraction

The void fraction is defined as the volumetric fraction of gas to the mixture of gas and liquid (see Fig. 9.2); thus,

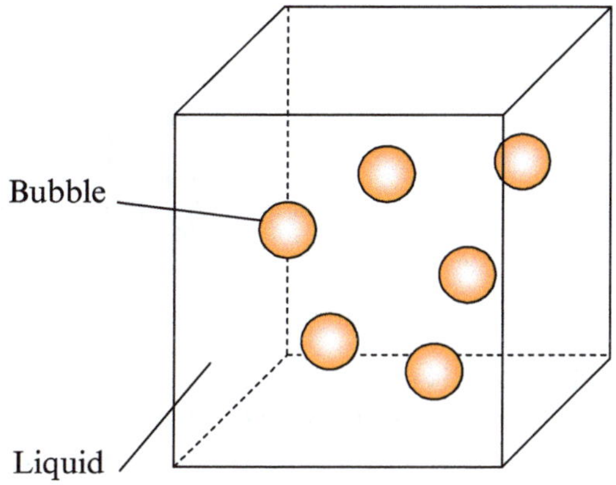

Fig. 9.2 A mixture of gas and liquid

$$\alpha_v = \frac{V_g}{V} \tag{9.1}$$

$$V = V_g + V_L \tag{9.2}$$

where V_g is the volume of gas (m^3), V_L is the volume of liquid (m^3), and V is the total volume of gas and liquid (m^3). This technical term is commonly used in the abovementioned engineering fields.

9.2.1.2 Flow Quality

The flow quality, χ, is defined as the ratio of the mass flow rate of gas (kg/s) to the total mass flow rate of a mixture of gas and liquid (kg/s):

$$\chi = \frac{\dot{m}_g}{\dot{m}} = \frac{\rho_g u_g A_g}{\rho_g u_g A_g + \rho_L u_L A_L} \cdots \quad (0 < \chi < 1) \tag{9.3}$$

$$\dot{m} = \dot{m}_g + \dot{m}_L \tag{9.4}$$

$$\dot{m}_g = \rho_g u_g A_g \tag{9.5}$$

$$\dot{m}_L = \rho_L u_L A_L \tag{9.6}$$

$$u_g = \frac{Q_g}{A_g}, \quad u_L = \frac{Q_L}{A_L} \tag{9.7}$$

where $\dot{m}_g$ is the mass flow rate of the gas (kg/s), $\dot{m}_L$ is the mass flow rate of the liquid (kg/s), $\dot{m}$ is the total mass flow rate of gas and liquid (kg/s), ρ_g is the density of gas, ρ_L is the density of liquid, u_g is the gas phase velocity, u_L is the liquid phase

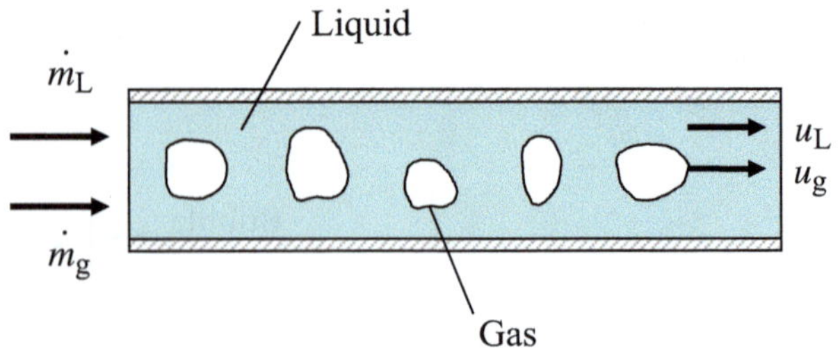

Fig. 9.3 Gas–liquid two-phase flow in a horizontal pipe

velocity, A_g is the cross-sectional area of the gas passage, A_L is the cross-sectional area of the liquid passage, Q_g is the gas flow rate, and Q_L is the liquid flow rate (see Fig. 9.3). The flow quality is mainly used in the analysis of gas–liquid two-phase flows in pipes and ducts.

9.2.1.3 Slip Ratio

The slip ratio, S (−), is defined as the ratio of the gas phase velocity, u_g (m/s), to the liquid phase velocity, u_L (m/s):

$$S = \frac{u_g}{u_L} \tag{9.8}$$

For homogeneous flow Eq. (9.8) reduces to

$$S = 1 \tag{9.9}$$

This technical term is used in every engineering field.

9.2.1.4 Relative Velocity

The relative velocity between the gas and liquid, u_r (m/s), is defined as

$$u_r = u_g - u_L \tag{9.10}$$

This term also is used in every engineering field.

9.2.1.5 Phase Velocity

The gas phase and liquid phase velocities are rewritten here to emphasize their importance in practical applications:

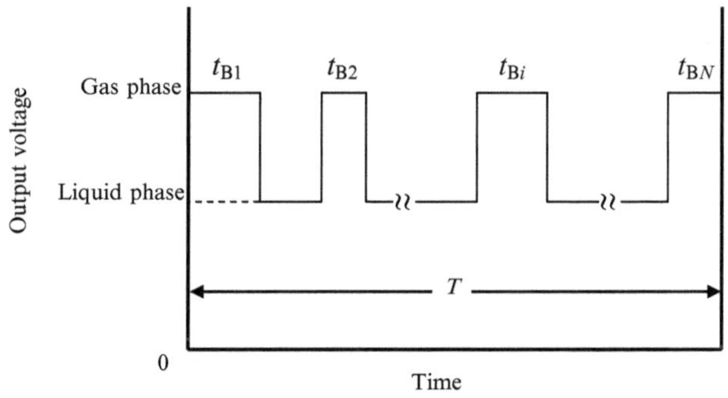

Fig. 9.4 Definition of gas holdup

$$u_g = \frac{Q_g}{A_g}, \quad u_L = \frac{Q_L}{A_L} \tag{9.11}$$

9.2.1.6 Superficial Velocity

The superficial velocities of gas and liquid, j_g (m/s) and j_L (m/s), are defined as

$$j_g = \frac{Q_g}{A}, \quad j_L = \frac{Q_L}{A} \tag{9.12}$$

$$j = j_g + j_L \tag{9.13}$$

where j is the total volumetric flux (m^3/m^2 s = m/s). These superficial velocities are imaginary velocities extensively used in chemical engineering.

9.2.1.7 Gas Holdup

Gas holdup, α, is defined as the ratio of the summation of bubble residence time at a point of measurement to the total duration of measurement, T (see Fig. 9.4). This parameter is widely used in materials and chemical engineering, as shown subsequently in more detail.

$$\alpha = \frac{\sum t_{Bi}}{T} \tag{9.14}$$

9.2.1.8 Bubble Frequency, f_B

The number of bubbles passing through a measurement point in one second is called the bubble frequency.

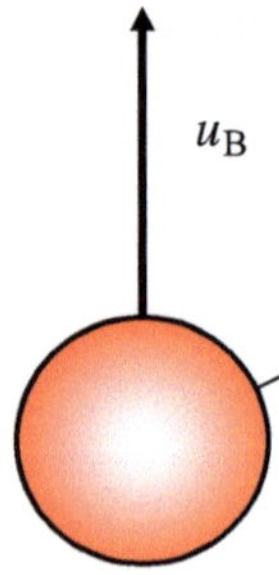

Fig. 9.5 Velocity of a rising bubble

9.2.1.9 Bubble Rising Velocity, u_B

The velocity of a bubble rising in a liquid bath or a vertical pipe is called the bubble rising velocity (see Fig. 9.5).

Exercise 9.1 Consider the situation in which gas is uniformly dispersed as small bubbles in a liquid flowing in a pipe. The liquid and bubbles are assumed to move at the same velocity. Such a flow is called homogeneous. As $u_g = u_L = u$ for the homogeneous flow, Eqs. (9.1) and (9.13) reduce to

$$\alpha_v = j_g/j \tag{9.15}$$

$$j = u \tag{9.16}$$

(1) Derive Eq. (9.16).
(2) Derive Eq. (9.15).

Answer to Exercise 9.1

1. The gas and liquid flow rates, Q_g and Q_L, are expressed by

$$Q_g = u_g A_g = u A_g \tag{9.17}$$

$$Q_L = u_L A_L = u A_L \tag{9.18}$$

The total volumetric flux, j, is given by

$$\begin{aligned} j = j_g + j_L = \frac{Q_g}{A} + \frac{Q_L}{A} = \frac{1}{A}(Q_g + Q_L) = \frac{1}{A}(uA_g + uA_L) \\ = \frac{u}{A}(A_g + A_L) = \frac{u}{A}A = u \end{aligned} \tag{9.19}$$

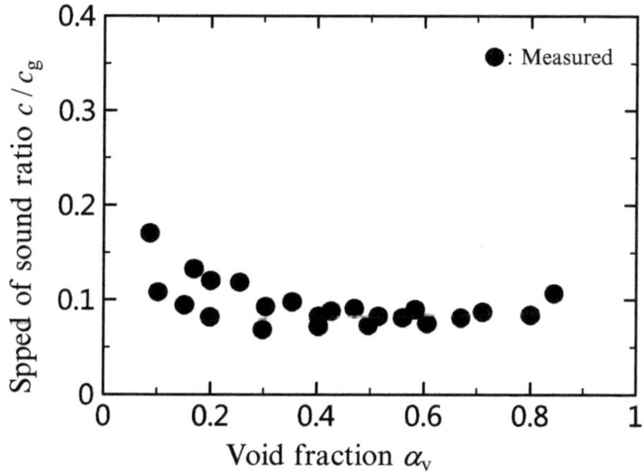

Fig. 9.6 Speed of sound in a mixture of gas and liquid

2. Consider a longitudinal section of the pipe of length, L. The time, T_g, required for a bubble to pass through this section is expressed by

$$T_{\mathrm{g}} = \frac{L}{u_{\mathrm{g}}} = \frac{L}{u} \tag{9.20}$$

The void fraction, α_v, is given by

$$\alpha_{\mathrm{v}} = \frac{V_{\mathrm{g}}}{V_{\mathrm{g}} + V_{\mathrm{L}}} = \frac{Q_{\mathrm{g}} T_{\mathrm{g}}}{AL} = \frac{Q_{\mathrm{g}} L}{uAL} = \frac{Q_{\mathrm{g}}/A}{u} = \frac{j_{\mathrm{g}}}{j} \tag{9.21}$$

9.2.2 Speed of Sound in a Mixture of Gas and Liquid

Propagation of pressure waves in a gas–liquid two-phase flow is strongly suppressed in the presence of dispersed bubbles. Figure 9.6 shows the relationship between the speed of sound, c, and the void fraction, α_v. It is evident that c is significantly affected by the void fraction. For example, the speed of sound in a single-phase water flow of approximately 1,500 m/s is drastically reduced to about 100 m/s, although the value of c in a single-phase air flow, denoted by c_g, is approximately 340 m/s. This information is essential in practical applications of gas–liquid two-phase flows [8].

9.3 Dynamical Models

The dynamical models proposed so far for gas–liquid two-phase flows can be classified into the following two categories:

9.3.1 Mixture Model

A mixture of gas and liquid is regarded as an imaginary fluid having mean physical properties of the gas and liquid. For example, the density of this model, ρ_m, is expressed by

$$\rho_m = \alpha_v \rho_g + (1 - \alpha_v)\rho_L \tag{9.22}$$

where α_v is the void fraction, ρ_g is the density of the gas, and ρ_L is the density of the liquid.

The previously introduced homogeneous flow is a typical example of this model. In addition, the slip model and drift flux model have been proposed [3, 8].

9.3.2 Two-Fluid Model

The governing equations for the motions of gas and liquid are coupled to predict the dynamic behaviors of gas–liquid two-phase flows. Discrimination of the gas and liquid phases and the momentum and mass exchanges between the two phases are key features to consider for obtaining reasonable results. A variety of algorithms have been proposed to address these features. Readers should refer to other texts [3] for further information.

9.4 Characteristics of Flow in Bottom-Blown Bath

9.4.1 Bubble Characteristics

The dispersion of bubbles in a bath agitated by gas injection can be characterized by the following quantities:

- Shape and size of bubbles
- Gas holdup, α
- Bubble frequency, f_B
- Mean bubble rising velocity, $\bar{u}_B$

The following sections discuss a bubbling jet generated in a cylindrical bath and a gas–liquid two-phase flow in a circular pipe.

9.4.1.1 Single Bubble in Still Liquid

The problem of a single bubble rising in a still liquid has been well investigated. Clift et al. [15] correlated the shape of a bubble as a function of the Morton number, *Mo*, the Eötvös number, *Eo*, and the bubble Reynolds number, *Re*, and presented some typical bubble shapes. When the volume of the bubble is very small, surface tension force is dominant and the bubble becomes spherical. As the bubble volume increases, the shape of the bubble deviates from spherical and a variety of nonspherical bubbles appear: ellipsoidal, wobbly, skirted, and cap-like. The relevant dimensionless numbers are defined thus:

$$Mo = \frac{g{\mu_L}^4 \Delta\rho}{{\rho_L}^2 \sigma^3} \tag{9.23}$$

$$Eo = \frac{g\Delta\rho {d_B}^2}{\sigma} \text{ (buoyancy force/force due to surface tension)} \tag{9.24}$$

$$Re = \frac{u_B d_B}{\nu_L} \text{ (inertial force/viscous force)} \tag{9.25}$$

where g is the acceleration due to gravity, μ_L is the dynamic viscosity of the liquid, $\Delta\rho = \rho_L - \rho_g$ is the density difference, ρ_L is the density of the liquid, ρ_g is the density of the gas, σ is the surface tension, d_B is the bubble diameter, u_B is the bubble rising velocity, and ν_L is the kinematic viscosity of the liquid.

9.4.1.2 Shape of Bubbles Generated at Single-Hole Nozzle in a Bath

Figure 9.7 shows that there are two types of bubble dispersion patterns in a bath agitated by bottom gas injection through a single-hole bottom nozzle: bubbling and jetting. The limit between the two patterns is described by the Mach number, *M*:

$$M = \frac{v_n}{c} = 1 \tag{9.26}$$

where v_n is the cross-sectional mean gas velocity at the nozzle exit and c is the speed of sound [16]. In the bubbling regime bubbles are successively generated at the nozzle exit and rise in the bath towards the bath surface due to the buoyancy forces acting on them. In the jetting regime a gas column is formed just above the nozzle which subsequently breaks into many small bubbles near the tip of the column.

The bubbling regime consists of bubbles being generated successively at the nozzle and, after a while, a recirculating flow is established in the bath, as shown in Fig. 9.8. The liquid in the bath ascends near the centerline, moves towards the side wall of the vessel, and descends along the side wall. Shear forces therefore act on

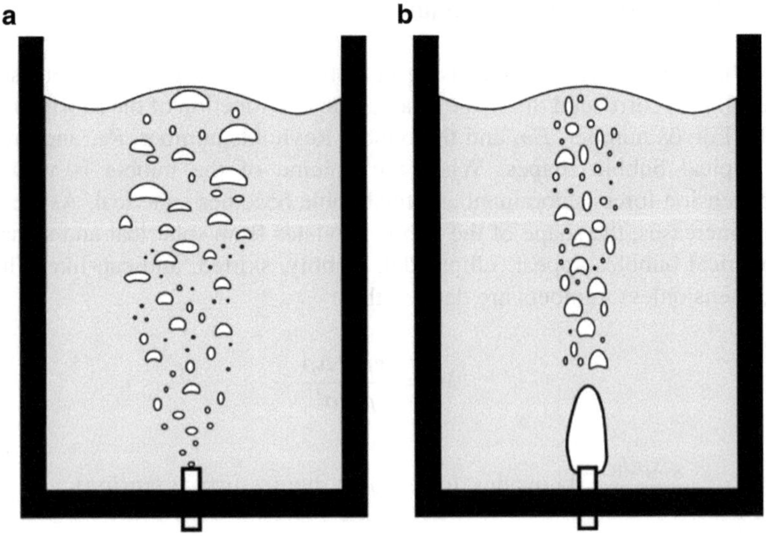

Fig. 9.7 Two types of bubble dispersion patterns in a bottom-blown bath. (**a**) Bubbling. (**b**) Jetting

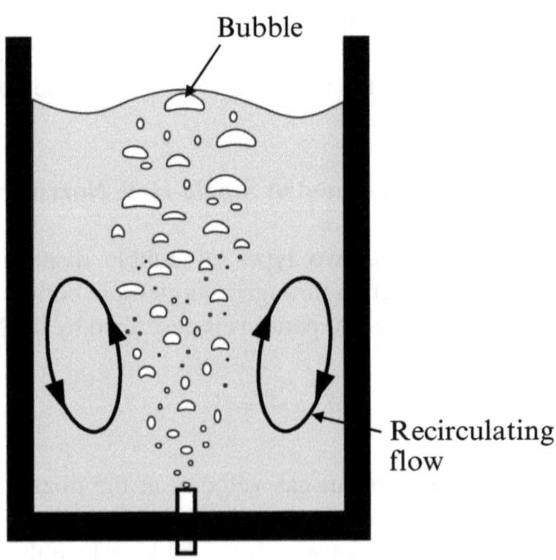

Fig. 9.8 Flow field in a bottom-blown bath

the bubbles in the course of rising in the bath. Accordingly, the shape of a bubble differs from that in a still liquid even if the bubble volume remains constant. In particular, when the bubble Reynolds number, Re_B, defined as

$$Re_B = \frac{\bar{u}_B \bar{d}_B}{\nu_f} \tag{9.27}$$

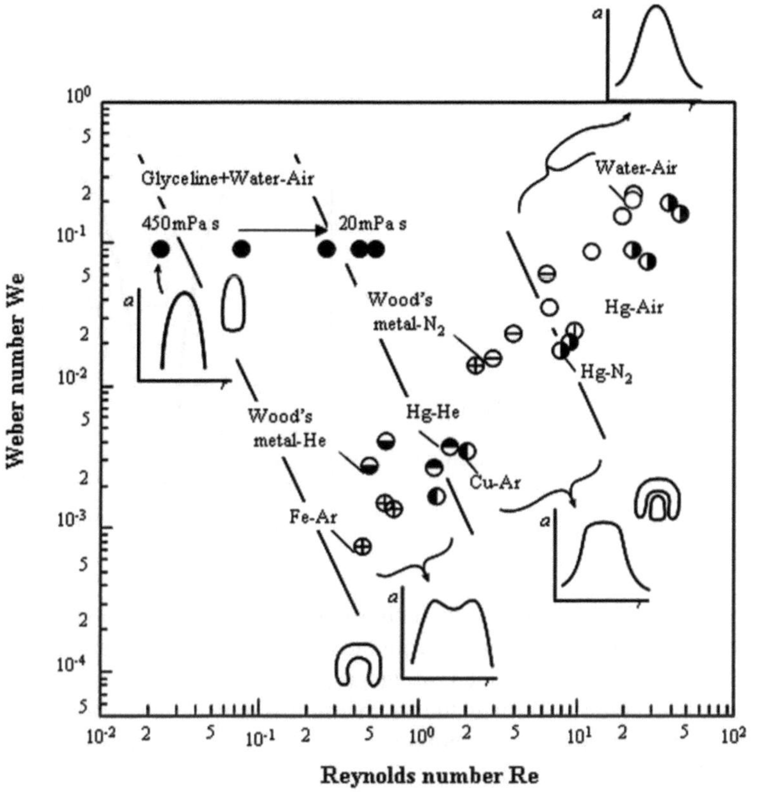

Fig. 9.9 Correlation of the shape of bubble generated successively in a bottom-blown bath

exceeds a critical value of approximately 400, the wake of each bubble undergoes transition to turbulence [17] and, as a result, every bubble is affected by the wake of the preceding bubble.

Previous experiments using electroresistivity probe [18, 19] have shown that the shape of a bubble can be correlated as a function of the modified Weber number, *We*, and the modified Reynolds number, *Re*, as shown in Fig. 9.9. It is interesting to note that a pair of bubbles is observed with one bubble being captured in the wake of the preceding one. Such a pair of bubbles is called a doublet. Evaluation of the interfacial area between the bubbles and liquid is complex, although it is closely associated with metallurgical reactions between the bubbles and the liquid. The Weber number, *We*, and the Reynolds number, *Re*, are expressed by

$$We = (\rho_g/\rho_L)\rho_L\bar{d}_B\bar{u}_B{}^2/\sigma \tag{9.28}$$

$$Re = (\rho_g/\rho_L)\bar{d}_B\bar{u}_B/\nu_L \tag{9.29}$$

where ρ_g is the density of the gas and ν_L is the kinematic viscosity of the liquid.

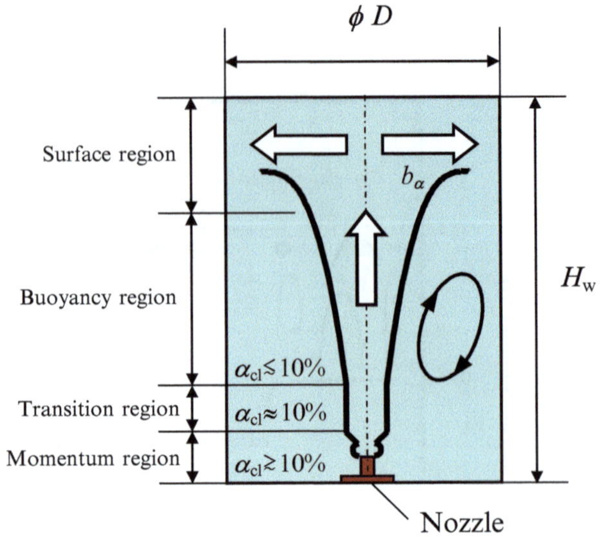

Fig. 9.10 Four fluid flow regions in a bottom-blown bath

The flow field in a bottom-blown bath is classified into four regions with respect to the vertical distance, z, measured from the nozzle exit, as schematically shown in Fig. 9.10. The inertial force of the issuing gas plays a significant role near the nozzle exit. Thus this region is termed the momentum region. In the buoyancy region, buoyancy forces acting on the bubbles are dominant. The upward liquid flow induced by rising bubbles changes direction at the bath surface towards the side wall and flows below the bath surface. This region therefore is termed the surface region. In the transition region, the dominant force changes from the inertial force of gas to the buoyancy forces acting on the rising bubbles.

Exercise 9.2 Air is injected into a water bath through a single-hole bottom nozzle at a flow rate, Q_g, of 200 cm^3/s. The inner diameter of the nozzle, d_{ni}, is 2.0 mm. Calculate the Mach number, M, in the nozzle. The speed of sound, c, is assumed to be 340 m/s.

Answer to Exercise 9.2 The cross-sectional mean air velocity at the nozzle exit, v_n, is given by

$$\begin{aligned} v_n &= 4Q_g/(\pi d_{ni}{}^2) = 4 \times 200/[\pi \times (0.20)^2] \\ &= 6.37 \times 10^3\,\text{cm/s} = 63.7\,\text{m/s} \end{aligned} \tag{9.30}$$

Substituting this value and $c = 340$ m/s into Eq. (9.26) yields

$$M = 0.19 \tag{9.31}$$

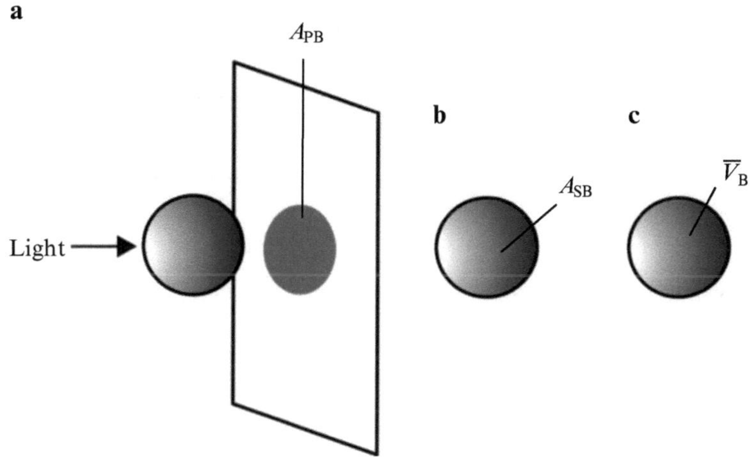

Fig. 9.11 Projected area A_{PB}, surface area A_{SB}, and volume of a solid particle $\overline{V}_B$

9.4.1.3 Bubble Size

Bubbles of varying sizes are often dispersed in a reactor and in a pipe as long as the flows are turbulent. Three quantities are relevant: characteristic diameter of a single bubble; probability distribution of the size of bubbles; mean bubble diameter. These quantities will be discussed in more detail below.

Characteristic Diameter of a Single Bubble

The geometrical diameter and the diameter of an equivalent sphere are usually used as the representative diameter of a single bubble [20–24]. The latter includes the following varieties (see Fig. 9.11):

(a) Projected equivalent diameter

$$\left(\frac{4A_{pB}}{\pi}\right)^{1/2} \tag{9.32}$$

where A_{pB} is the projected area of the bubble.

(b) Equivalent surface diameter

$$\left(\frac{A_{sB}}{\pi}\right)^{1/2} \tag{9.33}$$

where A_{sB} is the surface area of the bubble.

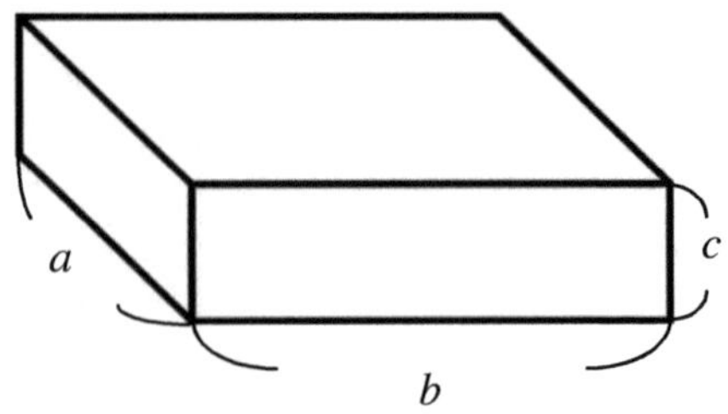

Fig. 9.12 Rectangular prism

(c) Equivalent volume diameter

$$\left(\frac{6V_B}{\pi}\right)^{1/3} \tag{9.34}$$

where V_B is the volume of the bubble.

Exercise 9.3 A rectangular prism has side lengths of a, b, and c, as shown in Fig. 9.12. Calculate the equivalent surface diameter and the equivalent volume diameter.

Answer to Exercise 9.3 The surface area, A_{sB}, of the prism is $2(ab + bc + ca)$, and the volume, V_B, is abc.

$$\text{Equivalent surface diameter} = (A_{sB}/\pi)^{1/2} = [2(ab + bc + ca)/\pi]^{1/2} \tag{9.35}$$

$$\text{Equivalent volume diameter} = (6V_B/\pi)^{1/3} = [6abc/\pi]^{1/3} \tag{9.36}$$

Probability Distribution

The normal distribution, lognormal distribution, and Weibull distribution are commonly used as the probability distribution of the size of bubbles:

(a) Normal distribution

$$y = \left[\frac{1}{(2\pi)^{1/2}\sigma}\right] \exp\left(-\frac{(x-\mu)^2}{2\sigma^2}\right) \tag{9.37}$$

where μ the mean value is finite and σ the standard deviation is positive.

(b) Lognormal distribution

$$y = \left[\frac{1}{(2\pi)^{1/2}\sigma x}\right] \exp\left[-\frac{1}{2}\left(\frac{(\log x - \mu)}{\sigma}\right)^2\right] \tag{9.38}$$

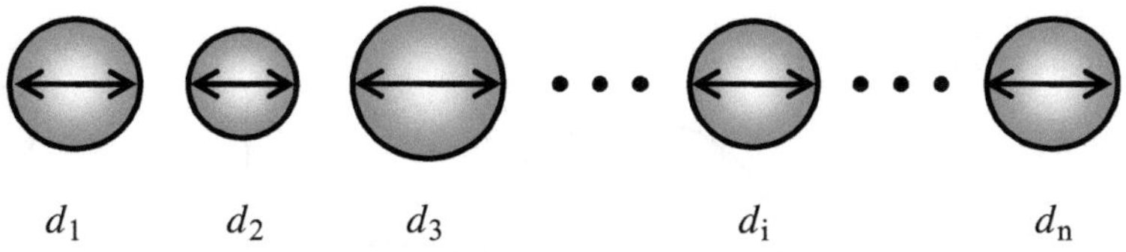

Fig. 9.13 Bubbles of different diameters

(c) Weibull distribution

$$y = \alpha\lambda x^{(\alpha-1)} \exp(-\lambda x^{\alpha}) \tag{9.39}$$

where $x \geq 0$, $\alpha > 0$, and $\lambda > 0$. In Eq. (9.39) α and λ are constant.

Mean Diameter

The following varieties of mean diameters have been proposed for bubbles [23–25] (see Fig. 9.13):

(a) Length mean diameter

$$D_{10} = \frac{\sum n_i d_i}{\sum n_i} \tag{9.40}$$

(b) Surface mean diameter

$$D_{20} = \left[\frac{\sum n_i {d_i}^2}{\sum n_i}\right]^{1/2} \tag{9.41}$$

(c) Volume mean diameter

$$D_{30} = \left[\frac{\sum n_i {d_i}^3}{\sum n_i}\right]^{1/3} \tag{9.42}$$

(d) Surface–length mean diameter

$$D_{21} = \frac{\sum n_i {d_i}^2}{\sum n_i d_i} \tag{9.43}$$

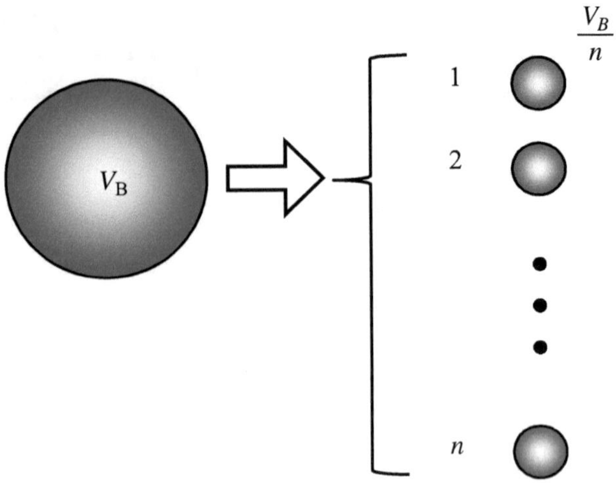

Fig. 9.14 Disintegration of a bubble into many small bubbles

(e) Volume–length mean diameter

$$D_{31} = \left[\frac{\sum n_i d_i^3}{\sum n_i d_i}\right]^{1/2} \tag{9.44}$$

(f) Volume–surface mean diameter

$$D_{32} = \frac{\sum n_i d_i^3}{\sum n_i d_i^2} \tag{9.45}$$

Mean Bubble Chord Length

It is difficult to measure the mean diameter of bubbles moving in an opaque liquid by means of still camera or high-speed video camera. Electroresistivity probe therefore is usually employed to measure bubble characteristics under this condition. The mean bubble chord length is often used instead of the mean bubble diameter [26].

Exercise 9.4 Consider a large bubble of volume V_B in volume that disintegrates into n smaller bubbles of equal volume (see Fig. 9.14). Derive the total surface area of the bubbles, A_{Bst}, as a function of n.

Answer to Exercise 9.4 The volume of each small bubble is V_B/n. Its diameter, d_{Bs}, and surface area, A_{Bs}, are expressed as follows:

$$d_{Bs} = [6V_B/(\pi n)]^{1/3} \tag{9.46}$$

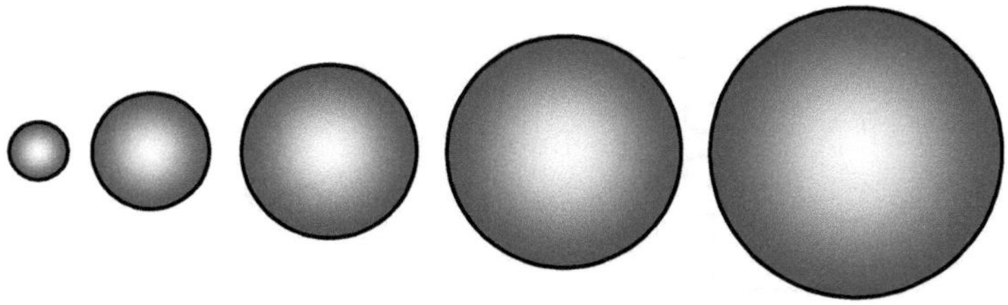

Fig. 9.15 Five spheres of different diameters

$$A_{\mathrm{Bs}} = \pi d_{\mathrm{Bs}}{}^2 = \pi[6V_{\mathrm{B}}/(\pi n)]^{2/3} \tag{9.47}$$

Accordingly, the total surface area is given by

$$\begin{aligned} A_{\mathrm{Bst}} &= \pi n[6V_{\mathrm{B}}/(\pi n)]^{2/3} \\ &= \pi[6V_{\mathrm{B}}/\pi]^{2/3} n^{1/3} \end{aligned} \tag{9.48}$$

Exercise 9.5 Consider five spheres of diameters 1, 2, 3, 4, and 5 cm (see Fig. 9.15). Calculate the abovementioned mean diameters, D_{10} through D_{32}.

Answer to Exercise 9.5 $D_{10} = 3$ cm, $D_{20} = 3.317$ cm, $D_{30} = 3.557$ cm, $D_{21} = 3.667$ cm, $D_{31} = 3.87$ cm, and $D_{32} = 4.091$ cm

9.4.1.4 Frequency of Bubble Formation in Bottom-Blown Bubbling Jet

Bubbles are formed under two types of conditions: constant pressure and constant flow [3, 16]. Constant flow condition is realized when the volume, V_c, of the gas chamber placed upstream of the nozzle is very small and the pressure loss in the gas delivery system is much greater than that associated with bubble formation (see Fig. 9.16a). Under this condition the flow rate of gas issuing out of the nozzle is kept constant. On the other hand, constant pressure condition is satisfied when V_c is very large (see Fig. 9.16b). The pressure in the nozzle exit is kept constant with time.

The wettability of the nozzle plays a significant role in determining the frequency of bubble formation at a single-hole bottom nozzle (see Fig. 9.17).

Gas injection is usually carried out under constant flow condition in real steelmaking processes to prevent weeping phenomenon. Thus, the following discussion will focus on the frequency of bubble formation under this condition.

Davidson and Amick proposed the following empirical relation for a wetted single-hole nozzle (see Fig. 9.17a) [3]:

$$f_{\mathrm{BF}} = 12.1 Q_{\mathrm{g}}{}^{0.133} d_{\mathrm{ni}}{}^{-0.434} \tag{9.49}$$

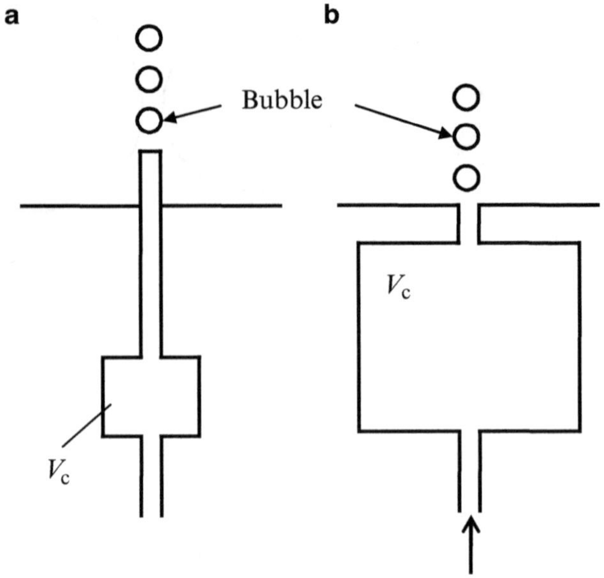

Fig. 9.16 Two types of bubble generation patterns from a single-hole nozzle. (**a**) Constant flow. (**b**) Constant pressure

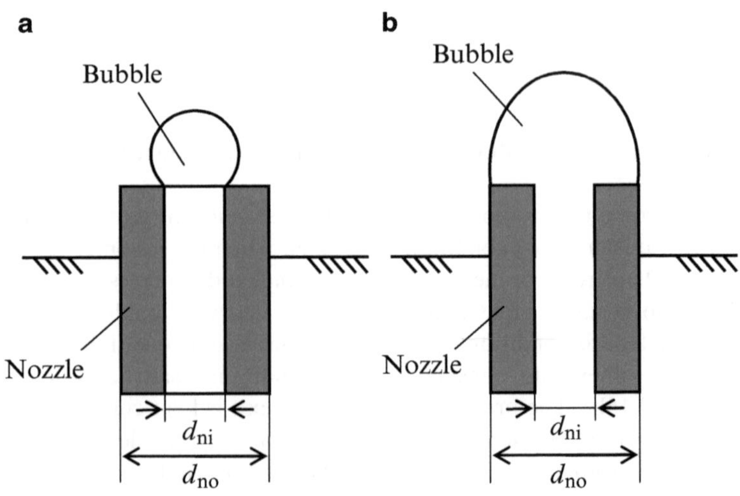

Fig. 9.17 Effect of nozzle wettability on the frequency of bubble formation from a single-hole nozzle under the constant flow condition. (**a**) Wetted nozzle. (**b**) Poorly wetted nozzle

where Q_g is the gas flow rate and d_{ni} is the inner diameter of the nozzle. The units of f_{BF}, Q_g, and d_{ni} are (Hz), (cm^3/s), and (cm), respectively. This equation was originally derived for air–water system.

Mori and Sano [16] later proposed the following empirical relation for f_{BF} for a poorly wetted single-hole nozzle with reference to Eq. (9.49) (see Fig. 9.17b):

$$f_{BF} = 12.1 {Q_g}^{0.133} {d_{no}}^{-0.434} \tag{9.50}$$

where d_{no} is the outer diameter of the nozzle. A comparison of Eq. (9.49) with Eq. (9.50) indicates that the frequency of bubble formation at a wetted nozzle is determined largely by the inner nozzle diameter, d_{ni}, while that at a poorly wetted nozzle is determined by the outer nozzle diameter, d_{no}. Mori and Sano demonstrated that Eq. (9.50) is applicable to a gas-molten metal system when the gas flow rate, Q_g, is relatively small.

The physical properties of fluids (gas and liquid) are not considered in the above equations. Ilegbusi et al. [3] therefore proposed the following empirical relation for a wetted nozzle:

$$f_{BF} = 1.06 (\rho_L g^3/\sigma)^{1/4} (\rho_g/\rho_L)^{1/5} \left[({Q_g}^2/g)^{1/5} / d_{ni} \right]^{1/2} \tag{9.51}$$

where ρ_L is the density of liquid, g is the acceleration due to gravity, σ is the surface tension, and ρ_g is the density of the gas. The SI unit should be used in calculating Eq. (9.51).

Equation (9.51) is modified for a poorly wetted nozzle as follows:

$$f_{BF} = 1.06 (\rho_L g^3/\sigma)^{1/4} (\rho_g/\rho_L)^{1/5} \left[({Q_g}^2/g)^{1/5} / d_{no} \right]^{1/2} \tag{9.52}$$

Exercise 9.6 Estimate the frequency of bubble formation, f_{BF} (Hz), and bubble diameter, d_B, for a wetted single-hole nozzle of inner diameter, d_{ni}, of 2.0 mm and outer diameter of 5.0 mm. The gas flow rate, Q_g, is 30 cm^3/s (see Fig. 9.18).

Answer to Exercise 9.6 The frequency of bubble formation can be calculated from Eq. (9.49) as follows:

$$\begin{aligned} f_{BF} &= 12.1 {Q_g}^{0.133} {d_{ni}}^{-0.434} = 12.1 (30)^{0.133} \times (0.20)^{-0.434} \\ &= 12.1 \times 1.572 \times 2.01 = 38.2\,\text{Hz} \end{aligned} \tag{9.53}$$

The gas flow rate, Q_g, is expressed by

$$Q_g = V_B f_{BF} = (\pi {d_B}^3/6) f_{BF} \tag{9.54}$$

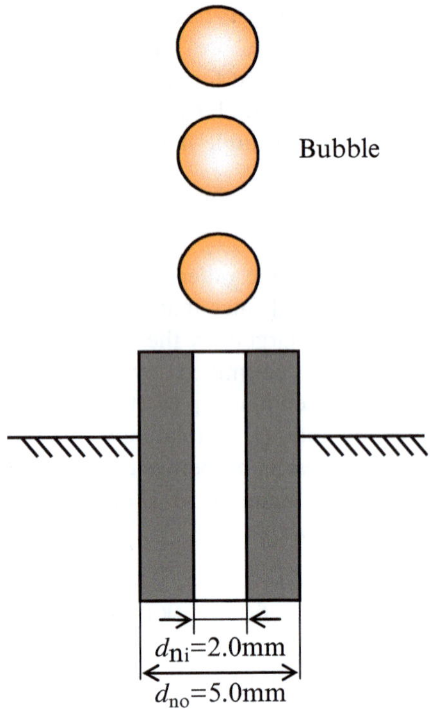

Fig. 9.18 Bubble formation from a wetted single-hole nozzle

The bubble diameter therefore is given by

$$\begin{aligned} d_B &= [6Q_g/(\pi f_{BF})]^{1/3} = 0.54[Q_g d_{ni}{}^{0.5}]^{0.289} \\ &= 0.54[30 \times (0.20)^{0.5}]^{0.289} = 1.14\,\text{cm} \end{aligned} \tag{9.55}$$

Exercise 9.7 Estimate the frequency of bubble formation, f_{BF} (Hz), and bubble diameter, d_B (cm), for a wetted single-hole nozzle of an inner diameter, d_{ni}, of 2.0 mm and an outer diameter of 5.0 mm on the basis of Eq. (9.51). The gas flow rate, Q_g, is 30 cm^3/s.

Answer to Exercise 9.7 The frequency of bubble formation can be calculated from Eq. (9.51) as follows:

$$\begin{aligned} f_{BF} &= 1.06(\rho_L g^3/\sigma)^{1/4}(\rho_g/\rho_L)^{1/5}\left[(Q_g{}^2/g)^{1/5}/d_{ni}\right]^{1/2} \\ &= 1.06 \times [997 \times 9.80^3/(73 \times 10^{-3})]^{1/4} \times (1.22/997)^{1/5} \\ &\quad \times [\{(30 \times 10^{-6})^2/9.80\}^{1/5}/(2.0 \times 10^{-3})]^{1/2} \\ &= 1.06 \times 59.9 \times 0.262 \times 2.22 = 36.9\,\text{Hz} \end{aligned} \tag{9.56}$$

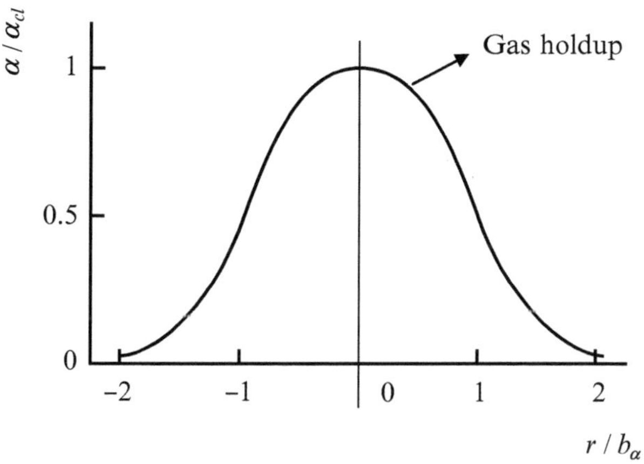

Fig. 9.19 Radial distribution of gas holdup in a bubbling jet

The bubble diameter, d_B, therefore is given by

$$d_B = [6Q_g/(\pi f_{BF})]^{1/3} = [6 \times 30 \times 10^{-6}/(3.14 \times 36.9)]^{1/3}$$
$$= 1.16 \times 10^{-2}\,\text{m} = 1.16\,\text{cm} \quad (9.57)$$

This value compares favorably with that calculated from Eq. (9.49).

9.4.1.5 Gas Holdup, α (%), in Bottom-Blown Bubbling Jet

Gas holdup, α (%), is defined as the ratio of the summation of bubble residence time at the tip of the electrode of an electroresistivity probe to total measurement time, T, [27, 28].

$$\alpha = 100 \sum t_{Bi}/T \quad (9.58)$$

where t_{Bi} is the residence time of the ith bubble at the measurement position.

The gas holdup is sometimes called the temporal void fraction. The radial distribution of gas holdup in a cylindrical bath agitated by centric bottom gas injection is expressed by the following normal (Gaussian) distribution (see Fig. 9.19):

$$\alpha = \alpha_{cl} \exp[-\ln 2 \cdot r^2/b_\alpha{}^2] \quad (9.59)$$

where α_{cl} is the centerline value of α, r is the radial distance, and b_α is the half-value radius of the gas holdup distribution. The details of α_{cl} and b_α can be found elsewhere [3].

Empirical equations for α_{cl} and b_α will be given below.

Momentum Region

$$\alpha_{\mathrm{cl}} = 50(z/z_0)^{-2n} \tag{9.60}$$

$$b_\alpha = b_\alpha(z_0)(z/z_0)^n \tag{9.61}$$

$$z_0 = 5.0 d_{\mathrm{ni}} Fr_{\mathrm{m}}{}^{0.3} \tag{9.62}$$

$$b_\alpha(z_0) = 0.42\left(Q_{\mathrm{g}}{}^2/g\right)^{1/5} \tag{9.63}$$

$$Fr_{\mathrm{m}} = \rho_{\mathrm{g}} Q_{\mathrm{g}}{}^2/(\rho_{\mathrm{L}} g d_{\mathrm{ni}}{}^5) \tag{9.64}$$

$$n = 0.0338 \log Fr_{\mathrm{m}} + 0.505 \ (1 < Fr_{\mathrm{m}} < 6.07 \times 10^4) \tag{9.65a}$$

$$= 2/3 \ (6.07 \times 10^4 < Fr_{\mathrm{m}} < 2 \times 10^5) \tag{9.65b}$$

Buoyancy Region

$$\alpha_{\mathrm{cl}} = 23 \exp(-1.5 z/z_5) \tag{9.66}$$

$$b_\alpha = 0.47 b_\alpha(z_5) \exp(0.75 z/z_5) \tag{9.67}$$

$$z_5 = 17\left(Q_{\mathrm{g}}{}^2/g\right)^{1/5} \tag{9.68}$$

$$b_\alpha(z_5) = 1.6\left(Q_{\mathrm{g}}{}^2/g\right)^{1/5} \tag{9.69}$$

where Fr_{m} is the modified Froude number and z_0 and z_5 denote the axial positions at which the gas holdup values are 50 and 5 %, respectively. The measured values of α_{cl} and b_α are satisfactorily predicted by these empirical equations. Information on empirical equations for other liquids should be referred to [3].

Exercise 9.8 Describe the physical meaning of the modified Froude number, Fr_{m}.

Answer to Exercise 9.8 The modified Froude number is rewritten as

$$\begin{aligned} Fr_{\mathrm{m}} &= \rho_{\mathrm{g}} Q_{\mathrm{g}}{}^2/(\rho_{\mathrm{L}} g d_{\mathrm{ni}}{}^5) \\ &\approx \rho_{\mathrm{g}} Q_{\mathrm{g}}[Q_{\mathrm{g}}/(\pi d_{\mathrm{ni}}{}^2/4)]/(\rho_{\mathrm{L}} g d_{\mathrm{ni}}{}^3) \\ &= \rho_{\mathrm{g}} Q_{\mathrm{g}} v_{\mathrm{n}}/(\rho_{\mathrm{L}} g d_{\mathrm{ni}}{}^3) \end{aligned} \tag{9.70}$$

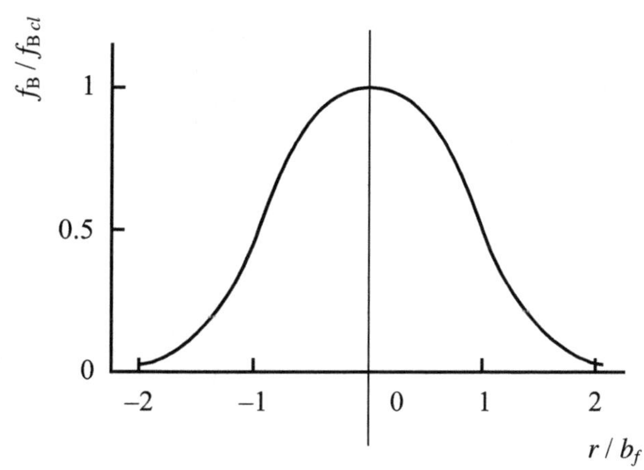

Fig. 9.20 Radial distribution of bubble frequency in a bubbling jet

where $\rho_g Q_g v_n$ denotes the inertial force of the injected gas. The volume of a bubble generated at the nozzle exit is assumed to be on the same order of magnitude as $d_{ni}{}^3$. In this case $\rho_L g d_{ni}{}^3$ denotes the buoyancy force acting on the bubble. Consequently, Fr_m implies the ratio of the inertial force of the injected gas to the buoyancy force acting on the bubble generated at the nozzle exit.

9.4.1.6 Bubble Frequency in Bottom-Blown Bubbling Jet

Bubble frequency, f_B, is defined as the number of bubbles passing through the electrode of an electroresistivity probe in 1 s. The radial distribution of the bubble frequency in a cylindrical bath agitated by centric bottom gas injection also is expressed by the following Gaussian distribution [3] (see Fig. 9.20):

$$f_B = f_{B,cl} \exp\left[-\ln 2 \cdot r^2/b_f{}^2\right] \tag{9.71}$$

where $f_{B,cl}$ is the centerline value of f_B, r is the radial distance, and b_f is the half-value radius of the bubble frequency distribution.

The half-value radius, b_f, is equal to b_α:

$$b_f = b_\alpha \tag{9.72}$$

Empirical relations for $f_{B,cl}$ are available in Ilegbusi et al. [3].

9.4.1.7 Mean Bubble Rising Velocity in Bottom-Blown Bubbling Jet

Attachment of surface active agents onto the bubble surface suppresses the surface renewal and, as a result, decreases the bubble rising velocity significantly.

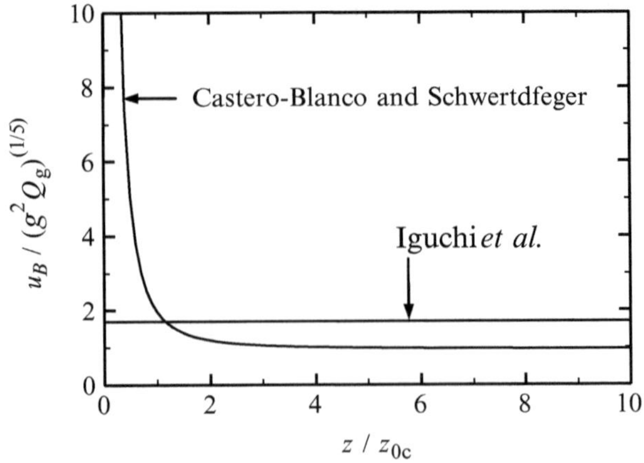

Fig. 9.21 Mean bubble rising velocity in a bubbling jet

In particular, a small bubble covered with the active agents behaves just like a solid body [29, 30]. Heat and mass transfer between gas and liquid also is affected by the agents. Careful attention should be given to the contamination of a liquid with surface active agents in carrying out water model experiments [31].

The radial distance r has minimal effect on the radial distribution of the mean bubble rising velocity in a bottom-blown bath. The following empirical relation is well established for the mean bubble rising velocity, $\bar{u}_{\mathrm{B}}$ [3] (see Fig. 9.21):

$$\bar{u}_{\mathrm{B}}/\left(g^2Q_{\mathrm{g}}\right)^{1/5} = 1.60(z/z_{0c})^{-2.04} + 1.82(z/z_{0c})^{-0.08} \quad 0.6 < z/z_{0c} \tag{9.73}$$

$$z_{0c} = 6.8d_{\mathrm{ni}}Fr_{\mathrm{m}}{}^{0.272} \tag{9.74}$$

where z_{0c} is the vertical location at which the gas holdup is 50 %. Ilegbusi et al. [3] derived the following simplified empirical relation:

$$\bar{u}_{\mathrm{B}} = 1.7\left(Q_{\mathrm{g}}g^2\right)^{1/5} \tag{9.75}$$

The measured values of gas holdup, α, and bubble frequency, f_{B}, follow the normal distribution, and the mean bubble rising velocity, $\bar{u}_{\mathrm{B}}$, is essentially independent of the radial distance, r, as evident in Figs. 9.22, 9.23, and 9.24.

Exercise 9.9 The radial distribution of gas holdup in a cylindrical bath agitated by centric bottom gas injection is expressed by Eq. (9.59). Show that the gas flow rate, Q_{g}, is calculated from the following relationship:

$$Q_{\mathrm{g}} = \int_0^R 2\pi r \bar{u}_{\mathrm{B}} \alpha \, dr \tag{9.76}$$

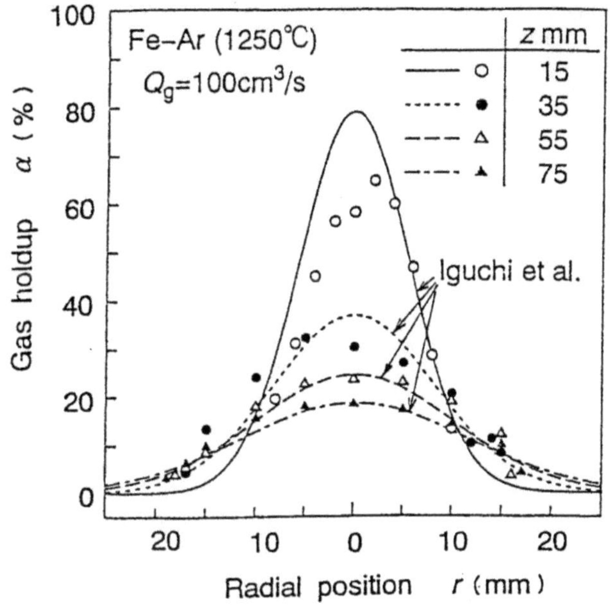

Fig. 9.22 Radial distribution of gas holdup in a molten iron bath agitated by bottom argon gas injection

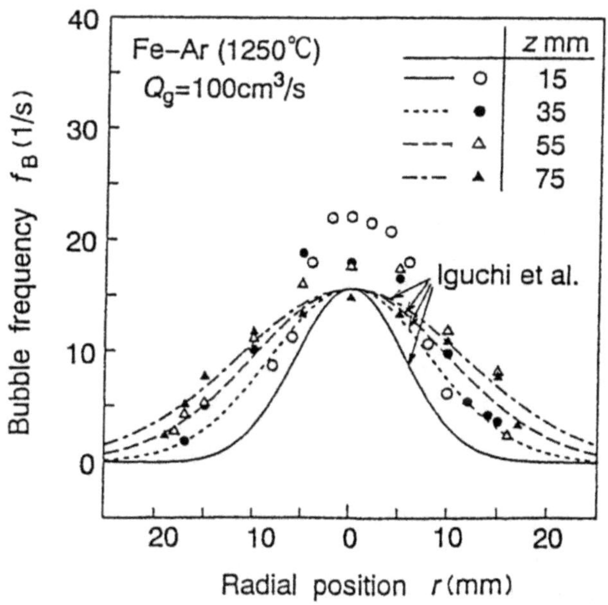

Fig. 9.23 Radial distribution of bubble frequency in a molten iron bath agitated by bottom argon gas injection

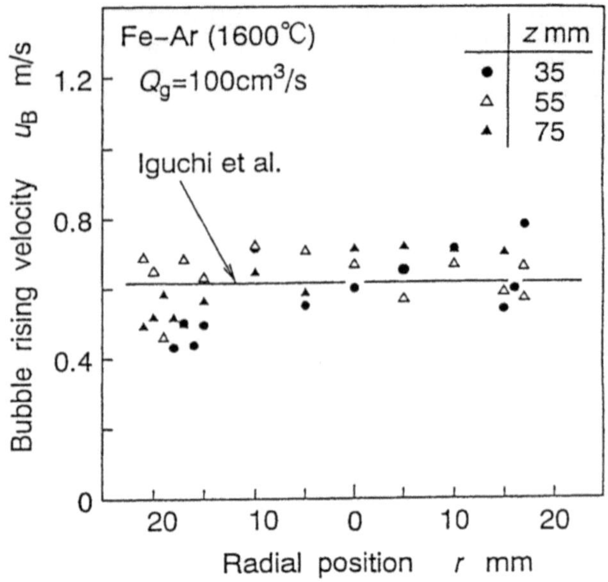

Fig. 9.24 Radial distribution of bubble rising velocity in a molten iron bath agitated by bottom argon gas injection

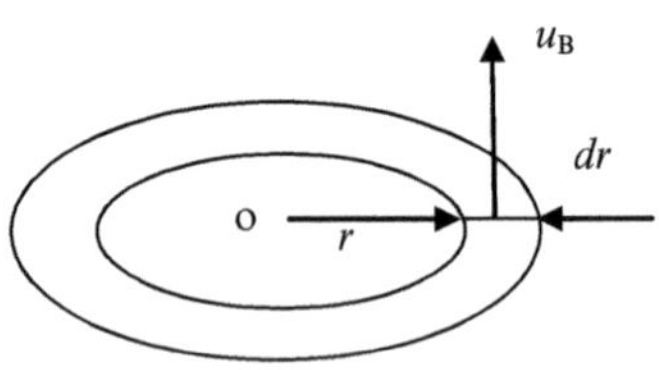

Fig. 9.25 Infinitesimal area element for calculating gas flow rate in a bath

Integrate Eq. (9.76) by assuming that the mean bubble rising velocity, $\bar{u}_B$, is constant.

Answer to Exercise 9.9 Figure 9.25 shows that the local gas flow rate, dQ_g, is given by

$$\begin{aligned} dQ_g &= (\text{gas flow rate}) \times (\text{gas holdup}) \\ &= \bar{u}_B dA \times \alpha \\ &= 2\pi r dr \bar{u}_B \alpha \end{aligned} \tag{9.77}$$

Integration of Eq. (9.77) gives Eq. (9.76).

Exercise 9.10 Air is injected into a cylindrical water bath through a centered single-hole bottom nozzle of inner diameter 2.0 mm. The nozzle is wetted by water. The gas flow rate, Q_g, is 300 cm^3/s. Calculate the mean bubble rising velocity, $\bar{u}_B$, at $z = 1.0$ m from Eq. (9.73).

Answer to Exercise 9.10 The modified Froude number, Fr_m, and z_{0c} are first calculated from Eqs. (9.70) and (9.74), respectively:

$$Fr_{\rm m} = \rho_{\rm g} Q_{\rm g}{}^2/(\rho_{\rm L} g d_{\rm ni}{}^5)$$
$$= 1.23 \times (300 \times 10^{-6})^2/[997 \times 9.80 \times (2.0 \times 10^{-3})^5] = 354 \qquad (9.78)$$

$$z_{0c} = 6.8 d_{\rm ni} Fr_{\rm m}{}^{0.272} = 6.8 \times 2.0 \times 10^{-3} \times (354)^{0.272} = 0.0671 \text{ m} \qquad (9.79)$$

As $z/z_{0c} = 1.0/0.0671 = 14.9$, the bubble rising velocity can be calculated from Eq. (9.73) as follows:

$$\bar{u}_{\rm B}/\left(g^2 Q_{\rm g}\right)^{1/5} = 1.60(z/z_{0c})^{-2.04} + 1.82(z/z_{0c})^{-0.08}$$
$$= 1.60 \times (14.9)^{-2.04} + 1.82 \times (14.9)^{-0.08}$$
$$= 1.47 \qquad (9.80)$$

Accordingly, $\bar{u}_{\rm B}$ is given by

$$\bar{u}_{\rm B} = 1.47\left(g^2 Q_{\rm g}\right)^{1/5} = 1.47 \times \left[(9.80)^2 \times (300 \times 10^{-6})\right]^{1/5}$$
$$= 0.723 \text{ m/s} \qquad (9.81)$$

9.4.2 Liquid Flow Characteristics: Velocity Distribution in Bottom-Blown Bath

Flow measurement in molten steel is difficult due to the high temperature and opacity of the fluid. Only the Karman vortex probe is available at present for measuring the meniscus velocity in the continuous casting mold. Thus, many cold model experiments have been conducted to understand molten steel flow characteristics in the steelmaking processes [32, 33]. Water is typically used to model molten steel because the kinematic viscosities of both fluids are nearly identical. Mercury and Wood's metal are also often used.

Figure 9.26 shows the visualized flow field in a water bath. There appears a large-scale ring vortex near the bath surface. The flow field was visualized by using aluminum foils floating in the bath. Figure 9.27 shows that the velocity distribution in the bubbling jet can be represented by the normal distribution just like the distribution of gas holdup, α, and bubble frequency, $f_{\rm B}$; thus,

$$\bar{u} = \bar{u}_{\rm cl} \exp\left[-\ln 2\left(\frac{r}{b_u}\right)^2\right] \qquad (9.82)$$

Castero-Blanco and Schwerdtfeger proposed the following empirical relation for the centerline velocity, $\bar{u}_{\rm cl}$, [3] (see Fig. 9.28).

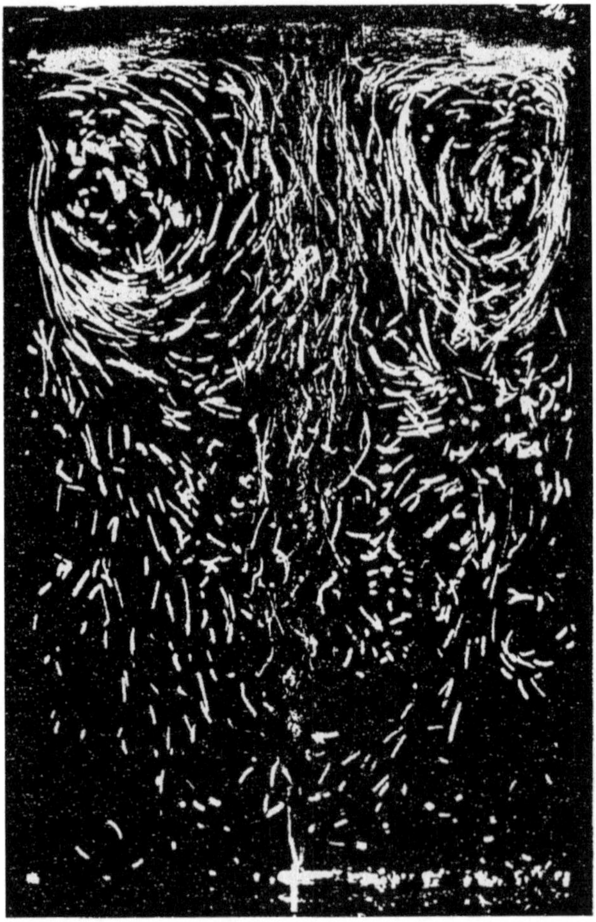

Fig. 9.26 Visualized flow field in a bottom-blown bath

$$\bar{u}_{cl}/\left(g^2Q_g\right)^{1/5} = 0.95(z/z_{0c})^{-2.30} + 1.10(z/z_{0c})^{-0.08}$$
$$0.8 < z/z_{0c} \quad (9.83)$$

Iguchi et al. [34] derived the following empirical relations:

$$\bar{u}_{cl} = 2.3u_r Fr_m{}^{0.036}\ (0.1 < P < 0.5) \quad (9.84)$$

$$= 1.2u_r P^{-0.28}\ (0.02 < P < 0.1) \quad (9.85)$$

$$u_r = (gQ_g/z)^{1/3} \quad (9.86)$$

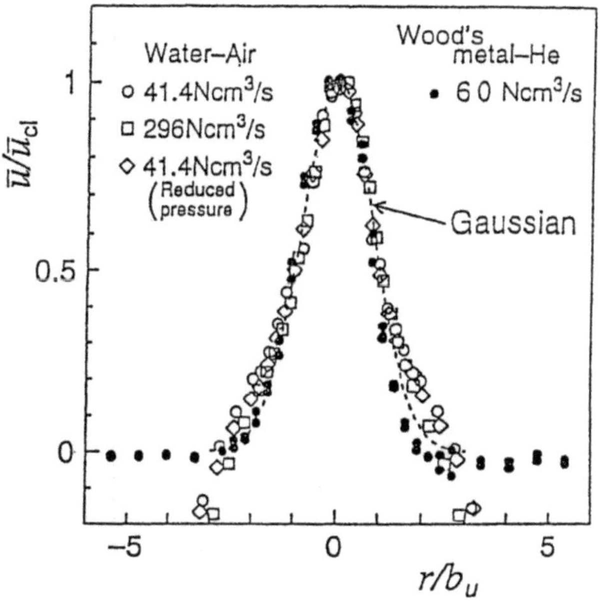

Fig. 9.27 Liquid velocity distribution in a bubbling jet caused by bottom gas injection

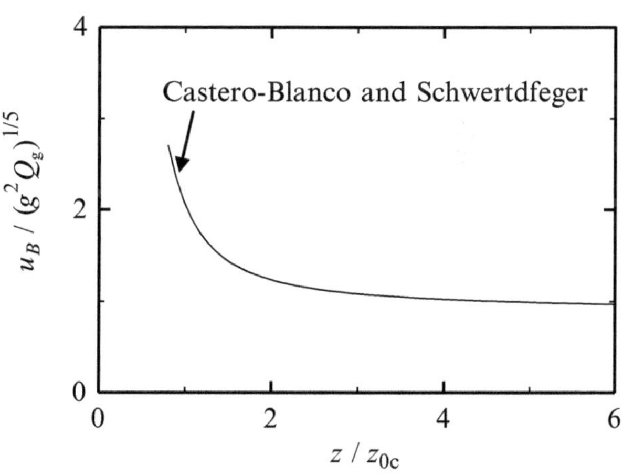

Fig. 9.28 Empirical equation for the liquid velocity on the centerline of a bubbling jet

$$P = \left[{Q_g}^2/\left(gz^5\right)\right]^{1/5} \tag{9.87}$$

$$b_u = 0.13z \tag{9.88}$$

where P is a kind of Froude number and b_u is the half-value radius of the distribution of $\bar{u}$.

Exercise 9.11 Air is injected into a cylindrical water bath through a centered single-hole bottom nozzle of inner diameter 2.0 mm (see Fig. 9.29). The nozzle is

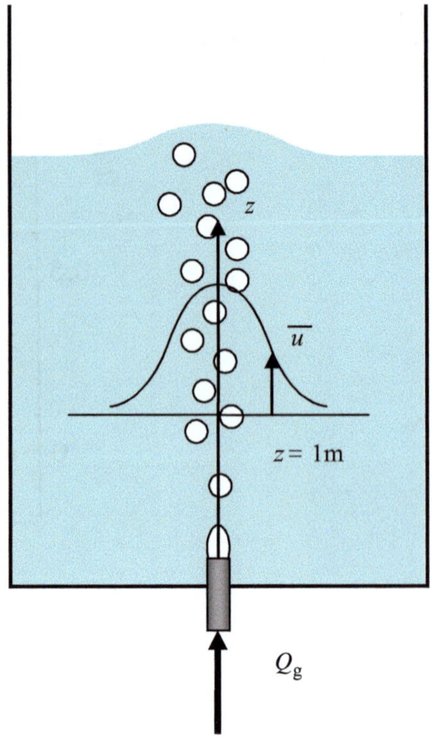

Fig. 9.29 Radial distribution of liquid velocity in a bubbling jet caused by bottom gas injection

wetted by water. The gas flow rate, Q_g, is 300 cm^3/s. Calculate the liquid velocity on the centerline of the bath, $\bar{u}_{cl}$, at $z = 1.0$ m from Eq. (9.83).

Answer to Exercise 9.11 The modified Froude number, Fr_m, and z_{0c} are already given in the previous Exercise 9.10 as

$$Fr_m = 354 \tag{9.89}$$

$$z_{0c} = 0.0671 \text{ m} \tag{9.90}$$

Since $z/z_{0c} = 1.0/0.0671 = 14.9$, the liquid velocity on the centerline of the bath can be calculated from Eq. (9.83) as follows:

$$\begin{aligned}\bar{u}_{cl}/\left(g^2 Q_g\right)^{1/5} &= 0.95(z/z_{0c})^{-2.30} + 1.10(z/z_{0c})^{-0.08} \\ &= 0.95 \times (14.9)^{-2.30} + 1.10 \times (14.9)^{-0.08} \\ &= 0.888\end{aligned} \tag{9.91}$$

Accordingly, $\bar{u}_{cl}$ is given by

$$\begin{aligned} \bar{u}_{cl} &= 0.888\left(g^2 Q_g\right)^{1/5} = 0.888 \times [(9.80)^2 \times (300 \times 10^{-6})]^{1/5} \\ &= 0.437\ \mathrm{m/s} \end{aligned} \tag{9.92}$$

Exercise 9.12 Air is injected into a cylindrical water bath through a centered single-hole bottom nozzle of inner diameter 2.0 mm. The nozzle is wetted by water. The gas flow rate, Q_g, is 300 cm^3/s. Calculate the liquid velocity on the centerline of the bath, $\bar{u}_{cl1}$, at $z = 1.0$ m from Eq. (9.84) or Eq. (9.85).

Answer to Exercise 9.12 The parameter P of Eq. (9.87) is

$$\begin{aligned} P &= \left[Q_g{}^2/\left(gz^5\right)\right]^{1/5} = [(300 \times 10^{-6})^2/(9.80 \times 1.0^5)]^{1/5} \\ &= 0.0246 \end{aligned} \tag{9.93}$$

The centerline velocity, $\bar{u}_{cl}$, is calculated from Eq. (9.85). In this equation u_r is given by

$$\begin{aligned} u_r &= \left(gQ_g/z\right)^{1/3} = (9.80 \times 300 \times 10^{-6}/1.0)^{1/3} \\ &= 0.143\ \mathrm{m/s} \end{aligned} \tag{9.94}$$

Substituting Eqs. (9.93) and (9.94) into Eq. (9.85) yields

$$\begin{aligned} \bar{u}_{cl} &= 1.2u_r P^{-0.28} = 1.2 \times 0.143 \times (0.0246)^{-0.28} \\ &= 0.485\ \mathrm{m/s} \end{aligned} \tag{9.95}$$

The calculated values in Eqs. (9.92) and (9.95) agree fairly well.

9.4.3 Measurement of Flow Phenomena

The methods of measuring bubble characteristics can be classified into two categories: visualization and electrical. The visualization method is suitable for grasping the whole flow field in a very short time. The dispersion of bubbles can readily be detected. In other words, the visualization method is effective for understanding the flow patterns of gas–liquid two-phase flows and unsteady flows. On the other hand, most of the electrical methods are regarded as point measurement technique. Therefore, it takes significant amount of time to get information on the whole flow field. Detailed information on the flow structure however can be obtained. Some of the conventionally employed methods will be briefly introduced in this section.

9.4.3.1 Still Camera

A still camera has extensively been used for measuring the shape and size of bubbles in a transparent bath. This method is effective when the number of bubbles is limited, and as a result, there is no overlap. The bubble rising velocity also can be obtained by dividing the displacement of bubble, ΔL, by a predetermined time increment, Δt. It is difficult to determine the bubble frequency and gas holdup by a still camera.

Many methods of analyzing overlapped bubbles have recently been proposed [35, 36]. Discrimination of each bubble from overlapped images is achievable using these new methods.

9.4.3.2 Video Camera

Home video camera: A home video camera records images at every 1/30 s. It can precisely detect the behavior of a bubble as long as the bubble rising velocity is low.

High-speed video camera: High-speed video cameras capable of taking images at 200–4,000 fps are widely used, for example, for investigating the generation and cessation of cavitation bubbles and bubble formation at a single-hole nozzle in a bottom-blown bath. Etoh et al. [37, 38] observed rupture phenomena of a single bubble at the bath surface using a 4,000 fps high-speed video camera and found two types of droplet generation mechanisms: jet drop and film drop [39; Etoh T, Private Communication]. The jet drop is generated when a small bubble arrives at the bath surface. Since the pressure in the bubble is much higher than that of the surrounding liquid, the gas in the bubble is ejected into the atmosphere just after it ruptures. This jet is called a micro jet. The liquid in the wake of the bubble is carried high into the atmosphere by the micro jet. On the other hand, the film droplets are induced by the rupture of a large bubble. The critical bubble size between the two types of drop generation mechanisms is not yet known even for the air–water system.

The vicinity of a dirty pond often smells due to methane gas generation. Many dirty liquid droplets are formed in the atmosphere due to rupture of the methane bubbles and are subsequently carried by the wind to the environment.

9.4.3.3 Electroresistivity Probe

Single-needle electroresistivity probe: This type of sensor was originally developed in boiler engineering [40]. Since then it has been widely utilized in many engineering fields such as mechanical and materials engineering to understand the behavior of gas–liquid two-phase flows in a pipe [41–44]. Two electrodes insulated except for their tips are immersed, for example, in a bath. A counter-electrode is immersed at another place in the bath to complete an electrical circuit. The electrical resistance between one of the electrodes and the counter-electrode becomes very high when the former electrode tip is surrounded by a rising bubble. Accordingly, the output voltage of the circuit changes drastically during the passage of bubbles. This probe is applicable to the measurement of bubble characteristics even in opaque liquids such as molten steel. It should be noted that one-needle electroresistivity probe is capable of only gas holdup and bubble frequency.

Two-needle electroresistivity probe: Two-needle electroresistivity probe can detect bubble rising velocity and bubble chord length in addition to gas holdup and bubble frequency. Detection of argon bubbles in a molten iron bath at a temperature of 1,600 °C is possible at present [42]. The life of the electrode was approximately 15 min at 1,600 °C and about 2 h at 1,250 °C in laboratory scale model experiments [34, 43]. No attempt has been made to apply the electroresistivity probe to real refining processes.

Multi-needle electroresistivity probe: Sekoguchi et al. [44] developed a multi-needle electroresistivity probe consisting of 409 needles to measure water–air two-phase flow in a vertical circular pipe. This probe was successfully used to obtain the three-dimensional shape of a train rising bubbles.

Bottom injection system has been utilized in steelmaking processes in which bubbles are successively generated at the nozzle exit and rise upwards. A comblike electroresistivity probe is used to detect the vertical section of a bubble rising above the nozzle.

9.4.3.4 Capacitance Probe

Fu and Evans [45] developed a unique sensor capable of measuring the bubble frequency, size, and mean bubble rising velocity by measuring the capacitance between a fine wire and a bubble. The wire is coated with insulator to prevent reaction with the surrounding molten metal. This feature prolongs the life of the probe, but the measurement accuracy is less than that of an electroresistivity probe.

9.4.3.5 Laser Optical Sensor

Optical fiber was used to measure the gas holdup, α, bubble frequency, f_B, and mean bubble rising velocity, $\bar{u}_B$, in a transparent bath [46]. The principles used to measure α and f_B are the same as those of an electroresistivity probe. The mean bubble rising velocity can be detected by processing the Doppler signals reflected from the bubble surface.

9.4.3.6 X-Ray CT, Ultrasonic CT, and X-Ray Method

Details of these methods should be referred to studies listed in [47, 48].

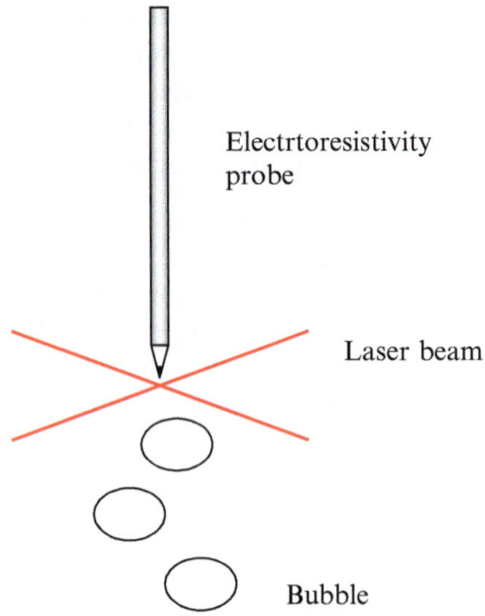

Fig. 9.30 Simultaneous measurement of liquid velocity and bubble rising velocity

9.4.3.7 Simultaneous Measurement of Liquid Velocity and Mean Bubble Rising Velocity

Consider the measurement of liquid velocity in a bubbling jet. The Doppler signals result not only from seeding particles but also from bubbles. Bubbles usually rise faster than the liquid around them due to buoyancy forces. Accordingly, it is necessary to discriminate the two signals. Sheng and Irons [32] placed a single-needle electroresistivity probe in the target volume of a one-dimensional laser Doppler velocimeter and succeeded in discriminating the Doppler signals from bubbles by monitoring the electroresistivity probe signal (see Fig. 9.30). Iguchi et al. [33] employed a two-dimensional LDV and a single-needle electroresistivity probe to discriminate the bubble signals. The root-mean-square values of the axial and radial turbulence components and the Reynolds shear stress of liquid flow were found to be largely unaffected by signals originating from bubbles as long as the gas holdup was less than approximately 30 %. The signals thus discriminated denote the velocity of the surface of the bubble.

9.4.3.8 Liquid Velocity Measurement Near Gas–Liquid Interface

Uemura and coworkers [49, 50] focused on a Hadamard-type small bubble appearing in a very low Reynolds number range and developed a unique method of visualizing the bubble and liquid interface. The velocity vectors of liquid around the interface were also obtained by particle tracking velocimetry (PTV). It is difficult at present to measure the velocity vectors in a bubble due to lack of adequate seeding particles for air flow.

9.4.3.9 Measurement of Liquid Film

Annular flow and mist–annular flow begin to appear with increased gas flow rate [51–53]. The surface behavior of such a liquid film and the liquid flow characteristics in the film are important in many engineering problems. Serizawa classified the methods of measuring liquid films into the following four categories [51, 52]:

(a) *Photographing method*: Direct photographing method and dye method.
(b) *Conductance method*: This method utilizes a change in the electrical resistance of liquid between two electrodes. Examples are the electroresistivity probe method, constant voltage method, and wall probe method.
(c) *Optical probe method*: Optical fiber probe, laser displacement sensor, and optical fiber liquid film sensor.
(d) *Ultrasonic method*: Ultrasonic echo method and ultrasonic velocimeter.

9.4.3.10 Flow Rate Measurement

The methods used to measure the flow rates of gas and liquid in gas–liquid two-phase flows have extensively been investigated mainly in atomic and petroleum engineering. The existing methods can be classified into the following two types [54].

(a) *Flow Rate Measurement Without Separating Gas and Liquid*
The Coriolis flow meter is applicable in situations in which the gas holdup distribution is uniform in a pipe. The total mass flow rate of gas and liquid can be obtained with satisfactory accuracy. Electromagnetic flow meter is applicable in the bubbly flow regime in a pipe. The total mass flow rate of gas and liquid can be measured by the constriction flow meter such as an orifice meter, a nozzle flow meter, and a Venturi flow meter. The accuracy of the constriction flow meters is typically low.
(b) *Individual Mass Flow Rate Measurement*
A turbine flow meter is suitable for this purpose. The accuracy is acceptable but not high.

9.4.3.11 Pressure Measurement

Pressure gauges developed for single-phase flows are basically applicable to gas–liquid two-phase flows as long as flash-type gauges are used [55]. When the pressure measurement position and the pressure transducer are connected with a small pipe, bubbles captured in the pipe would significantly affect pressure data especially in unsteady and high-speed flows. In addition, measurement of pressure

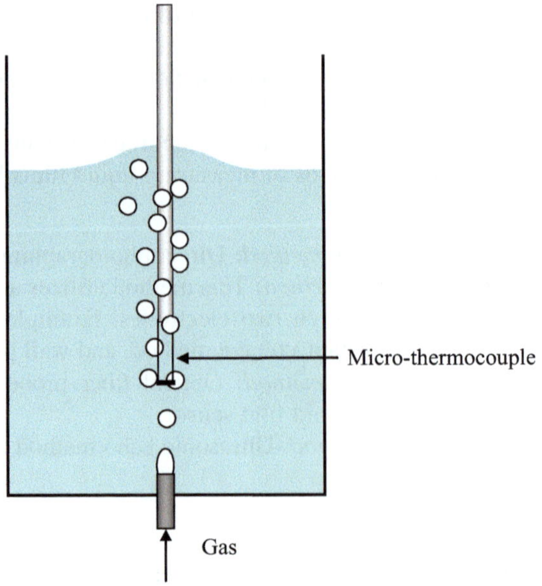

Fig. 9.31 Micro-thermocouple capable of measuring liquid and bubble temperatures

in pipeline systems accompanied by gas condensation and liquid evaporation is quite difficult.

9.4.3.12 Temperature Measurement

The gas and liquid phases pass a measurement zone alternately in gas–liquid two-phase flows, as shown in Fig. 9.31. Provided the gas and liquid have the same temperature, the temperature measurement methods developed for single-phase flows are applicable without any modification. However, when they have different temperature, care must be taken in the measurement. Iguchi et al. [56, 57] used a micro-thermocouple with junction diameter of 12.5 μm to simultaneously measure the gas and liquid temperatures in a bottom-blown bath. The temperature of the injected helium gas was much lower than that of the liquid in the bath. Precise measurement was possible only when the gas flow rate was high enough to blow a liquid droplet from the junction.

9.5 Heat and Mass Transfer

9.5.1 Heat Transfer from Sphere Immersed in Bubbling Jet

The mean Nusselt number, Nu_{pm}, is expressed as a function of the Reynolds number, Re, the Prandtl number, Pr, and the turbulence intensity, Tu. Iguchi et al. [58] demonstrated that the turbulence intensity plays an important role in the convection heat transfer around a solid sphere (see Fig. 9.32) and proposed the following relation based on Whitaker's equation [3]:

$$Nu_{pm} - 2 = \left(0.4{Re_p}^{1/2} + 0.06{Re_p}^{2/3}\right)Pr^{0.4}(\mu_L/\mu_0)^{1/4}(1+Tu)^{1.36}$$
$$(1,500 < Re_p < 16,000, Tu < 0.5, 5.7 < Pr < 9.2,$$
$$0.46 < (\mu_L/\mu_0) < 0.7) \quad (9.96)$$

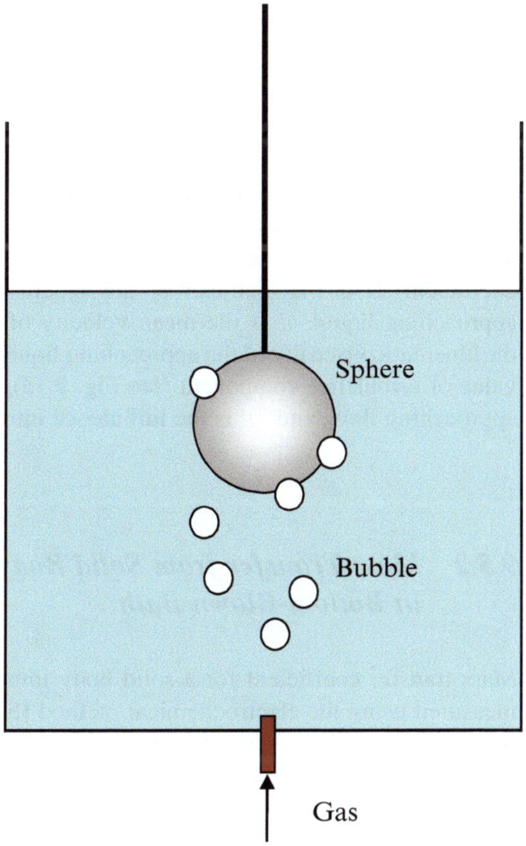

Fig. 9.32 Heat transfer from a sphere immersed in a bubbling jet

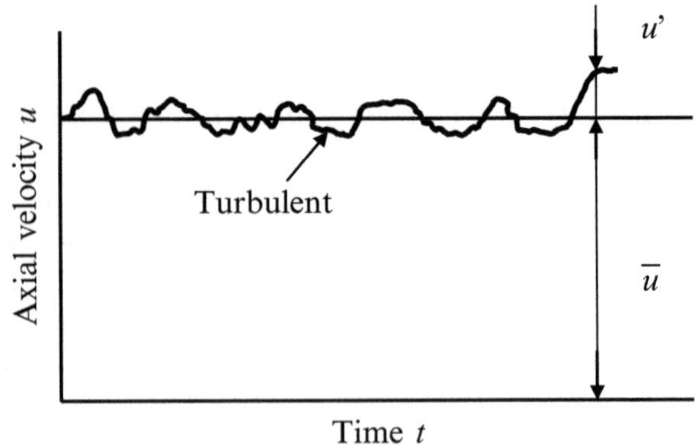

Fig. 9.33 Turbulent liquid velocity signal in a bubbling jet

$$Nu_{pm} = aD/\lambda_L \tag{9.97}$$

$$Re_p = \frac{\bar{u}D}{\nu_L} \tag{9.98}$$

$$Tu = \frac{u\prime_{rms}}{\bar{u}} \tag{9.99}$$

$$Pr = \nu_L/\kappa_L \tag{9.100}$$

where Re_p is the Reynolds number based on the diameter of the sphere, Pr is the Prandtl number, μ_L is the dynamic viscosity of the approaching flow, μ_0 is the dynamic viscosity of the fluid at solid surface temperature, α is the heat-transfer coefficient, D is the diameter of the sphere, λ_L is the thermal conductivity of approaching liquid, $\bar{u}$ is the mean velocity of the approaching liquid flow, ν_L is the kinematic viscosity of the approaching liquid flow, u'_{rms} is the root-mean-square value of turbulence component (see Fig. 9.33), κ_L is the thermal diffusivity of the approaching flow, and Tu is the turbulence intensity.

9.5.2 Mass Transfer from Solid Body Immersed in Bottom-Blown Bath

Mass transfer coefficient for a solid body immersed in a bottom-blown bath was measured using the electrochemical method [59]. It was found that analogy exists between heat and mass transfer in a flow field accompanied by turbulence intensity greater than 20 %. The results can be summarized as follows:

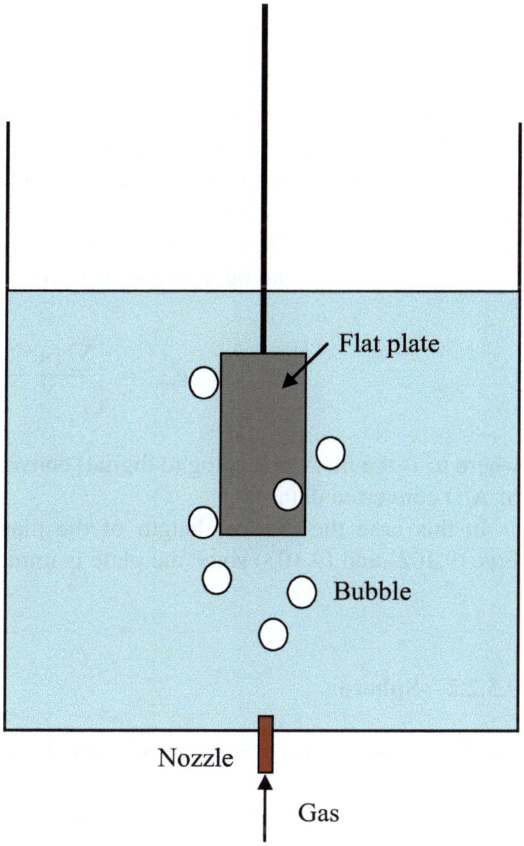

Fig. 9.34 Heat transfer from a flat plate immersed in a bubbling jet

9.5.2.1 Flat Plate

The mass transfer coefficient for a flat plate (see Fig. 9.34) can be correlated by the following empirical relation within a scatter of ±20 %:

$$Sh_{\mathrm{pm}} = 0.664 Re_{\mathrm{p}}{}^{(0.5+0.10Tu)} Sc^{1/3} (10^2 < Re_{\mathrm{p}} < 10^4, 0.4 < Tu < 5) \tag{9.101}$$

$$Sh_{\mathrm{pm}} = \frac{k_{\mathrm{m}} L}{D_i} \tag{9.102}$$

$$Re_{\mathrm{p}} = \frac{\bar{u} L}{\nu_{\mathrm{L}}} \tag{9.103}$$

$$Tu = \frac{u\prime_{\mathrm{rms}}}{\bar{u}} \tag{9.104}$$

$$Sc = \frac{\nu_L}{D_i} \tag{9.105}$$

where Sh_{pm} is the mean Sherwood number, Re_p is the Reynolds number, Sc is the Schmidt number, k_m is the mass transfer coefficient, L is the representative length, D_i is the diffusion coefficient, $\bar{u}$ is the mean velocity of approaching flow, ν_L is the kinematic viscosity of the liquid, u'_{rms} is the root-mean-square value of the turbulence component, and Tu is the turbulence intensity. The root-mean-square value of u', denoted by u'_{rms}, can be determined from

$$u'_{rms} = \left[\frac{\sum (u_i')^2}{N}\right]^{1/2} \tag{9.106}$$

where u_i' is the ith A/D (analog to digital) converted value of u' and N is the number of A/D converted data.

In this case the vertical length of the plate should be substituted into L in Eqs. (9.102) and (9.103) since the plate is immersed vertically in the bath.

9.5.2.2 Sphere

The following empirical relation is applicable to a sphere (see Fig. 9.35).

$$Sh_{pm} = 2 + 0.60 Re_p{}^{(0.5+0.10Tu)} Sc^{1/3} (10^2 < Re_p < 10^4, 0.3 < Tu < 0.5) \tag{9.107}$$

The diameter of the sphere, D, should be used as the representative length, L, in Eqs. (9.102) and (9.103). The existing measured values of the mass transfer coefficient from a sphere also can be predicted by Eq. (9.107) within a scatter of ±20 %.

9.5.2.3 Horizontal Circular Cylinder

The following empirical equation is applicable to a horizontal circular cylinder (see Fig. 9.36):

$$Sh_{pm} = 0.63 Re_p{}^{(0.5+0.10Tu)} Sc^{1/3} (10^2 < Re_p < 10^4, 0.3 < Tu < 0.5) \tag{9.108}$$

The diameter of the cylinder, D, should be used as the representative length, L, in Eqs. (9.102) and (9.104). Equation (9.108) can predict the measured values of Mazumdar et al. within a scatter of ±50 % [3].

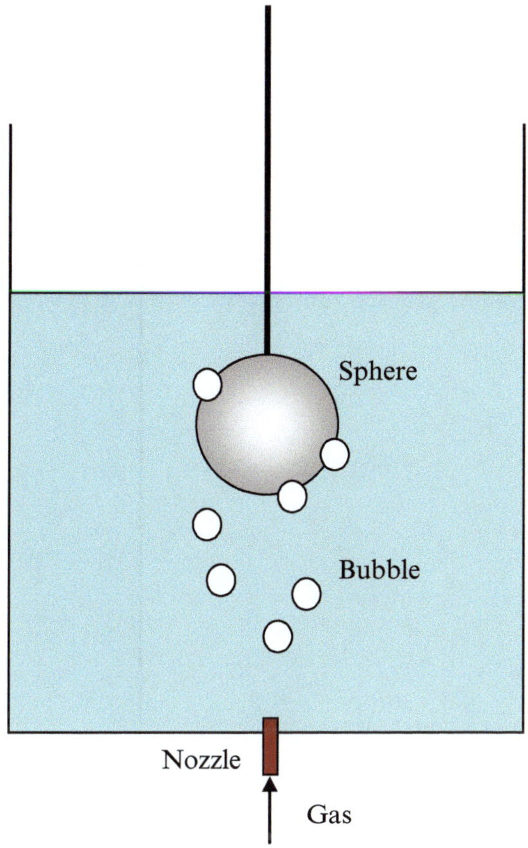

Fig. 9.35 Mass transfer from a sphere immersed in a bubbling jet

9.6 Dimensionless Parameters

Eötvös Number, Eo

$$Eo = g{d_B}^2 \Delta\rho/\sigma \tag{9.109}$$

where g is the acceleration due to gravity, d_B is the bubble diameter, $\Delta\rho = \rho_L - \rho_g$ is the density difference, and σ is the surface tension. This parameter represents the ratio of the buoyancy force acting on a bubble to the force due to surface tension.

Bubble Reynolds Number in Still Liquid

$$Re_B = u_B d_B / \nu_L \tag{9.110}$$

where u_B, d_B, and ν_L are the bubble rising velocity, bubble diameter, and kinematic viscosity of liquid, respectively. This dimensionless parameter expresses the ratio of the inertial force of bubble to the viscous force acting on the bubble.

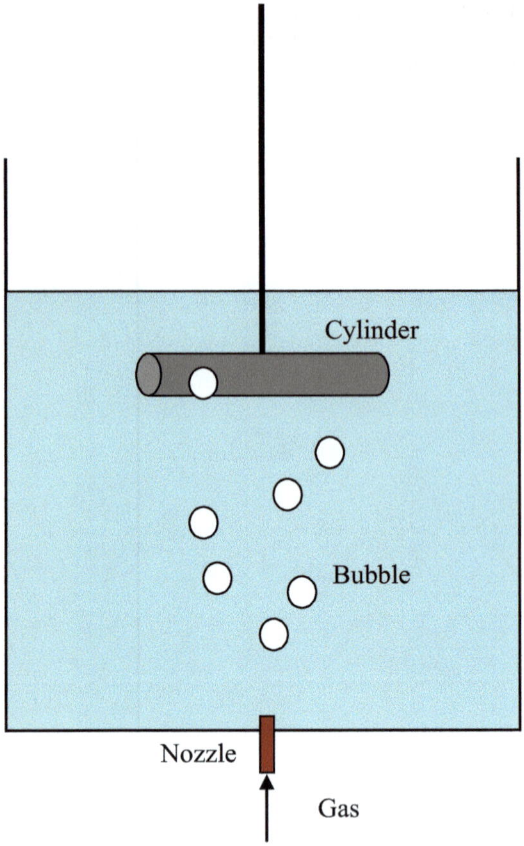

Fig. 9.36 Mass transfer from a horizontal circular cylinder immersed in a bubbling jet

Mach Number in Gas Injection System

$$M = v_n / c \tag{9.111}$$

where v_n and c are the cross-sectional mean velocity of gas in the nozzle and the speed of sound, respectively.

Weber Number in Gas Injection System

$$We = (\rho_g/\rho_L)\rho_L \bar{d}_B \bar{u}_B{}^2/\sigma \tag{9.112}$$

Reynolds Number in Gas Injection System

$$Re = (\rho_g/\rho_L)\bar{d}_B \bar{u}_B/\nu_L \tag{9.113}$$

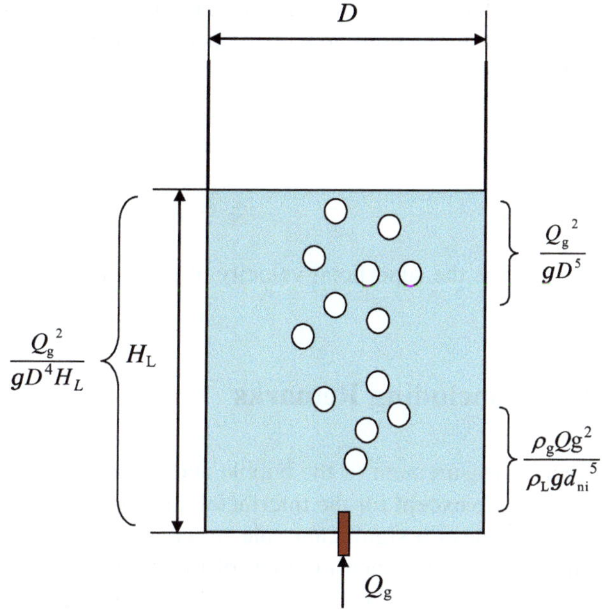

Fig. 9.37 Three kinds of modified Froude numbers applied to the flow field in a bottom-blown bath

Modified Froude Number in Gas Injection System 1

$$Fr_m = \rho_g Q_g^2/(\rho_L g d_{ni}^5) \tag{9.114}$$

This modified Froude number is useful for correlating the fluid flow phenomena near the nozzle exit (see Fig. 9.37).

Modified Froude Number in Gas Injection System 2

$$Fr_m = Q_g^2/(gD^5) \tag{9.115}$$

This parameter can be rewritten as

$$Fr_m = [(Q_g^2/g)^{1/5}/D]^5 \tag{9.116}$$

where $(Q_g^2/g)^{1/5}$ is a measure of the diameter of bubble dispersion region on the bath surface. This bubble dispersion region is sometimes called the plume eye. Accordingly, $(Q_g^2/g)^{1/5}/D$ expresses the ratio of the diameter of the bubble dispersion region to the vessel diameter. It is usually used for correlating the fluid flow phenomena near the bath surface (see Fig. 9.37).

Modified Froude Number in Gas Injection System 3

$$Fr_m = Q_g^2/(gD^4H_L) \tag{9.117}$$

The bath depth, H_L, is included in Eq. (9.117), and, hence, this parameter is used for correlating the fluid flow phenomena in the whole bath (see Fig. 9.37). Equation (9.117) is rewritten thus:

$$Fr_m \approx {v_{sp}}^2/gH_L \tag{9.118}$$

$$v_{sp} = Q_g/(\pi D^2/4) \tag{9.119}$$

where v_{sp} is the superficial velocity of gas in the bath.

9.7 Concluding Remarks

Precise measurement of the bubble and liquid flow characteristics in water models are possible except for the interfacial area between gas and liquid phases, although this area plays a significant role on chemical reactions between the two phases. Unfortunately, the measurement of these characteristics in real processes is quite difficult. There has been no reported success in the measurement of molten metal velocity even in single-phase molten steel flow, for the following reasons:

- Molten steel is at relatively high temperature of approximately 1,600 °C (1,873 K).
- Molten steel is opaque.
- Molten steel contains a variety of impurities and nonmetallic inclusions which are likely to react with velocity sensors.
- Liquid flow in molten steel bath is usually highly turbulent and unsteady. Erosion of sensors is a major problem under these conditions, making it imperative to develop novel and precise flow measuring techniques.

References

1. ISIJ (1984) 100th and 101st Nishiyama Memorial Lecture. ISIJ, Tokyo
2. Sahai Y, St. Pierre GR (1992) Advances in transport processes in metallurgical systems. Elsevier, Amsterdam
3. Ilegbusi OJ, Iguchi M, Wahnsiedler W (1999) Mathematical and physical modeling of materials processing operations. Chapman-Hall/CRC, Boca Raton
4. Kajioka H (1997) Ladle metallurgy. ISIJ, Tokyo
5. Mazumdar D, Guthrie RIL (1995) The physical and mathematical modeling of gas stirred ladle systems. ISIJ Int 35(1):1–20
6. Szekely J (1979) Fluid flow phenomena in metals processing. Academic, New York
7. JSPS 19th Committee (1994) Recent investigations on non-metallic inclusions in molten steel. JSPS, Tokyo
8. Iguchi M, Takei M, Matsui G (2008) Basic characteristics of fluid and thermal engineering. Asakura Book Co., Ltd., Tokyo

9. Terauchi Y, Iguchi M, Kosaka H, Yokoya S, Hara S (1999) Wettability effect on the flow pattern of air-water two-phase flows in a vertical circular pipe. Tetsu-to-Hagane 85 (9):645–651
10. Taniguchi S, Kikuchi A (1992) Mechanisms of collision and coagulation between fine particles in fluid. Tetsu-to-Hagane 78(4):527–535
11. ISIJ (1996) History of continuous casting of steel in Japan. ISIJ Tokyo, 511
12. Lin ZH, Guthrie RIL (1994) Modeling of metallurgical emulsions. Metall Mater Trans B 25B:855–864
13. Takashima S, Iguchi M (2000) Metal droplet holdup in the thick slag layer subjected to bottom gas injection. Tetsu-to-Hagane 86(4):217–224
14. ISIJ (1976) Modeling theory and Scale-up in steelmaking. ISIJ, Tokyo
15. Clift R, Grace JR, Weber ME (1978) Bubbles, drops, and particles. Academic, New York, 27
16. Mori K, Sano M (1981) Process kinetics in injection metallurgy. Tetsu-to-Hagane 67:672–695
17. Hetsroni G (1989) Particles-turbulence interaction. Int J Multiphase Flow 15:735–746
18. Iguchi M, Nakatani T, Kawabata H (1997) Development of a multi-needle electroresistivity probe for measuring bubble characteristics in molten metal baths. Metall Mater Trans B 28B:409–416
19. Iguchi M, Nakatani T, Tokunaga H (1997) The shape of bubbles rising near the nozzle exit in molten metal baths. Metall Mater Trans B 28B:417–423
20. Akagawa K (1980) Gas-liquid two-phase flow. Corona Book Co., Ltd., Tokyo
21. Ueda T (1989) Gas-liquid two-phase flow, fluid flow and heat transfer. Yokendo Book Co., Ltd., Tokyo
22. Hetsroni G (1982) Handbook of multiphase systems. McGraw-Hill, New York
23. Jinbo G et al(1989) Handbook of fine particles. Asakura Book Co., Ltd, Tokyo
24. Chem. Eng. Soc. Jpn. (1999) Handbook of chemical engineering, 6th edn. Tokyo, p 231
25. Jpn. Soc. Powder Engineering (1998) Handbook of powder engineering, 2nd edn. Nikkan Kogyo Shinbun Book Co., Ltd, Tokyo
26. Kawakami M, Tomomoto N, Itoh K (1982) Statistical analysis of gas bubbles dispersion in liquid phase – water model experiment on bottom blowing processes. Tetsu-to-Hagane 68 (7):774–783
27. Kasagi N et al(1997) Handbook of fluid flow experiments. Asakura Book Co., Ltd., Tokyo, p 600
28. Serizawa A (1974) Fluid dynamics characteristics of two-phase flow. Dr. Dissertation, Kyoto University, Kyoto
29. Clift R, Grace JR, Weber ME (1978) Bubbles, drops, and particles. Academic, New York, 171
30. Tomiyama A, Kataoka I, Sakaguchi T (1995) Drag coefficients of bubbles (1st report, drag coefficients of a single bubble in a stagnant liquid). Trans Jpn Soc Mech Eng Ser B 61 (587):2357–2364
31. Kumagai T, Iguchi M (2001) Effects of liquid flow on gas-liquid mass transfer. Tetsu-to-Hagane 87(4):159–166
32. Sheng Y, Irons GA (1991) A combined laser Doppler anemometry and electrical probe diagnostic for bubbly two-phase flow. Int J Multiphase Flow 17(5):585–598
33. Iguchi M, Kondoh T, Uemura T (1995) Simultaneous measurement of liquid and bubble velocities in a cylindrical bath subject to centric bottom gas injection. Int J Multiphase Flow 20:753–762
34. Iguchi M, Kawabata H, Morita Z, Nakajima K, Itoh Y (1994) Continuous measurements of bubble characteristics in a molten iron bath with Ar gas bubbling. Tetsu-to-Hagane 80 (5):365–370
35. Song X, Yamamoto F, Iguchi M, Shen L, Ruan X, Ishii K (1998) A method for counting particle number in overlapped particle images. ISIJ Int 38(9):971–976
36. Song X, Yamamoto F, Iguchi M (2001) Image processing (particle image velocimetry). Ferrum 6(4):250–258
37. Etoh T, Takehara K, Nakagawa K (1995) Rupture of a bubble. J Flow Meas 12(17):27

38. Etoh T, Takehara K (1995) Proceedings of the third International symposium on air-water gas transfer. Heidelberg, p 447
39. Uchida T, Iguchi M (2000) Entrapment of small bubbles into liquid by breakup of bubble escaping from bath surface. J Mater Process Manuf Sci 8(3):256–264
40. Akagawa K (1980) Gas-liquid two-phase flow. Corona Book Co., Ltd., Tokyo, p 119
41. Castellejos AH, Brimacombe JK (1989) Physical characteristics of gas jets injected vertically upward into liquid metal. Metall Mater Trans B 20:595–601
42. Iguchi M, Kawabata H, Nakajima K, Morita Z (1995) Measurement of bubble characteristics in a molten iron bath at 1600°C using an electro-resistivity probe. Metall Mater Trans B 26:67–74
43. Iguchi M, Kawabata H, Morita Z, Nakajima K (1996) Bubble characteristics in molten copper bath with gas injection. Trans Jpn Soc Mech Eng Ser B 62(593):79–84
44. Sekoguchi K, Takeishi M, Ishimatsu T (1985) Interfacial structure in vertical upward annular flow. PhysicoChemical Hydrodynamics 6(1/2):239
45. Fu Q, Evans JW (1999) A capacitance probe for measurement of bubbles in molten metals. ISIJ Int 39(4):309–318
46. Iguchi M, Kaji M, Okita K (1998) Development of a laser optical sensor for measuring bubble characteristics in an oily liquid bath. ISIJ Int 38:802–806
47. Iguchi M (2001) Flow visualization. Ferrum 6(3):180–184
48. Takenaka N (2001) Radiography for visualization and measurement of gas-liquid two-phase flow. J Visualization Soc Jpn 21:14–19
49. Uemura T, Yamauchi S (2001) PTV measurement in the vicinity of liquid-liquid two-phase boundary. Graphic Simulation of Multiphase Flow 51(3):1
50. Yamauchi S, Uemura T, Ozawa M (2000) PTV measurement at close region to interface of a water drop in oil. J Jpn Soc Multiphase Flow 14(4):466–472
51. Serizawa A (2000) Measurement techniques for film flow. J Jpn Soc Multiphase Flow 14(1):39–57
52. Akagawa K (1980) Gas-liquid two-phase flow. Corona Book Co., Ltd., Tokyo, p 135
53. Minemura K, Yamasita M (2000) A review on measuring methods of gas-liquid flow rates. J Jpn Soc Multiphase Flow 14(1):16–23
54. Sasaki Y, Iguchi M (2001) Flow rate measurement. Ferrum 6(2):84–90
55. Iguchi M, Matsuura K (2000) Pressure measurement. Ferrum 5(10):731–735
56. Iguchi M, Tokunaga H, Tatemichi H, Morita Z (1992) Heat transfer between bubbles and liquid during cold gas injection. Tetsu-to-Hagane 78(3):415–422
57. Tokunaga H, Iguchi M, Tatemichi H (1995) Heat transfer between bubbles and molten wood's metal. ISIJ Int 35(1):21–25
58. Iguchi M, Tani J, Uemura T, Morita Z (1989) Flow phenomena and heat transfer around a sphere submerged in water jet and bubbling jet. ISIJ Int 29(8):658–665
59. Iguchi M, Tomida H, Nakajima K, Morita Z (1993) Mass transfer from a solid body immersed in a cylindrical bath with bottom gas injection. ISIJ Int 33(7):728–734

Chapter 10
Flows Involving Solid Phase

Abstract Solid particle injection processes are often used in the steelmaking industry. Explanation is mainly given on solid–gas and solid–liquid two-phase flows in a pipe and a reactor together with solid–gas two-phase jet. Experimental techniques for the shape, size, number, and moving velocity of solid particles are discussed.

Keywords Coulter counter • Flow pattern • Fluidized bed • Particle image velocimetry • Particle injection • Particle-laden jet

10.1 Introduction

Flows involving solid phase abound in materials processing operations. Pulverized coal injection into blast furnace is a typical example of gas–solid two-phase flows in the ironmaking and steelmaking processes [1, 2]. The removal of nonmetallic inclusions such as alumina (Al_2O_3) from molten steel still remains a primary concern of researchers in continuous casting processes [3]. This is a typical example of solid–liquid two-phase flows of interest in this chapter.

Understanding the dynamic behaviors of bubbles and nonmetallic inclusions in real molten metal baths with temperature much higher than 1,000 °C is currently a challenge. Cold model experiments therefore have been carried out using water, mercury, and Wood's metal to simulate real molten metals [1–6]. This chapter describes the dynamic behaviors of solid particles in a fluid and the measurement of fluid flow induced by the particles. Fortunately, most of the measurement techniques are similar to those proposed for gas–liquid two-phase flows [7]. The essence of those methods will be recalled here. The most significant difference between gas–liquid and solid–liquid two-phase flows is that the gravitational force acting on solid particles is much larger than that on bubbles.

M. Iguchi and O.J. Ilegbusi, *Basic Transport Phenomena in Materials Engineering*,
DOI 10.1007/978-4-431-54020-5_10,

10.2 Dynamics of Solid Particles in Two-Phase Systems

10.2.1 Shape and Size of Particle

The shape and size of a particle do not change with respect to time and location, unlike a bubble. When many fine solid particles gather, they sometimes form clusters. This trend indicates that the description of a system involving many particles is quite complex. The shape and size of a particle can be described by the parameters discussed for a bubble in a previous Sect. 9.4.1.

10.2.2 Mixing Ratio and Concentration

10.2.2.1 Solid–Gas Two-Phase Flow in a Pipe

The following mixing ratio [8, 9] is commonly used for describing a solid–gas two-phase flow in a pipe (see Fig. 10.1):

$$\text{Mixing ratio} = \frac{\dot{m}_\text{s}}{\dot{m}_\text{g}} \tag{10.1}$$

where $\dot{m}_\text{s}$ and $\dot{m}_\text{g}$ are the mass flow rates of solid and gas, respectively. The ratio of the volumetric flow rate of solid particles, Q_s, to that of gas, Q_g, is called the concentration [10]:

$$\text{Concentration} = \frac{Q_\text{s}}{Q_\text{g}} \tag{10.2}$$

10.2.2.2 Solid–Liquid Two-Phase Flow in a Pipe

The concentration of solid particles in a solid–liquid two-phase flow can be described by the following volumetric concentration, C_v, and mass concentration, C_m:

$$C_\text{v} = \frac{Q_\text{s}}{Q_\text{L} + Q_\text{s}} \tag{10.3}$$

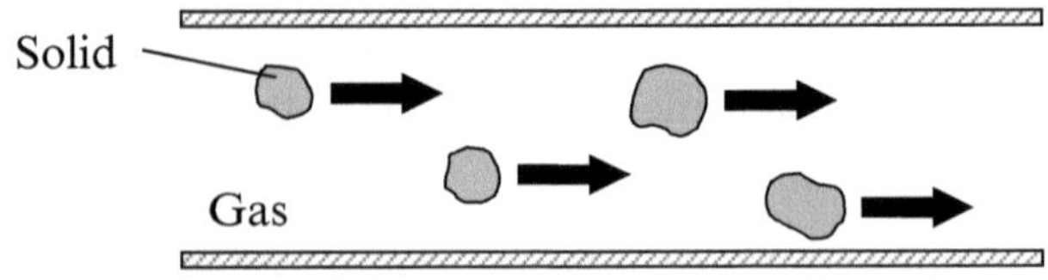

Fig. 10.1 Gas–solid two-phase flow in a pipe

$$C_{\mathrm{m}} = \frac{\dot{m}_{\mathrm{s}}}{\dot{m}_{\mathrm{L}} + \dot{m}_{\mathrm{s}}} \tag{10.4}$$

where Q_L and $\dot{m}_L$ are the volumetric flow rate and mass flow rate of the liquid, respectively. Although a technical term called weight concentration is often used, it is equal to the mass concentration.

The volumetric ratio of solid particles in a mixture of solid particles and gas also is used to describe the concentration of solid particles [8]. Therefore, care must be taken to the definition of concentration in treating two-phase flows containing solid particles.

10.2.3 Flow Pattern of Solid–Gas Two-Phase Flows

Flow pattern implies the spatial and temporal dispersion structure of solid particles in a system under consideration [10]. From a particle interaction point of view, the flow patterns of solid–gas two-phase flows can be classified into the following three categories [11, 12] (see Fig. 10.2):

Type (1): Collision-free flow
Type (2): Collision dominated flow
Type (3): Contact dominated flow

The flow pattern can be characterized by the velocity of the gas phase and the particle concentration. The collision-free flow appears for high gas velocity and low particle concentration. With a decrease in the gas velocity and an increase in the particle concentration, the flow pattern shifts from Type (2) to Type (3)

10.2.3.1 Solid–Gas Two-Phase Flows in a Horizontal Pipe

The flow pattern varies basically from uniform to stratified via circulating flow and slug flow with increase in the number of solid particles (see Fig. 10.3). In real pipeline systems, the flow patterns of solid–gas two-phase flows are very complex compared to those of gas–liquid two-phase flows. This is because the mutual attachment of solid particles, interparticle friction, electrostatic force acting, and so on, play a complex role on the behavior of the particles. The flow pattern can be shown as a function of the pressure gradient and the superficial velocity of gas [9–12].

10.2.3.2 Solid–Gas Two-Phase Flows in a Vertical Pipe

The flow pattern depends strongly on the pressure gradient and the superficial velocity of gas in the same manner as that in solid–gas two-phase flow in a vertical pipe (see Fig. 10.4). The uniform flow, plug flow, and choking flow are commonly observed [12].

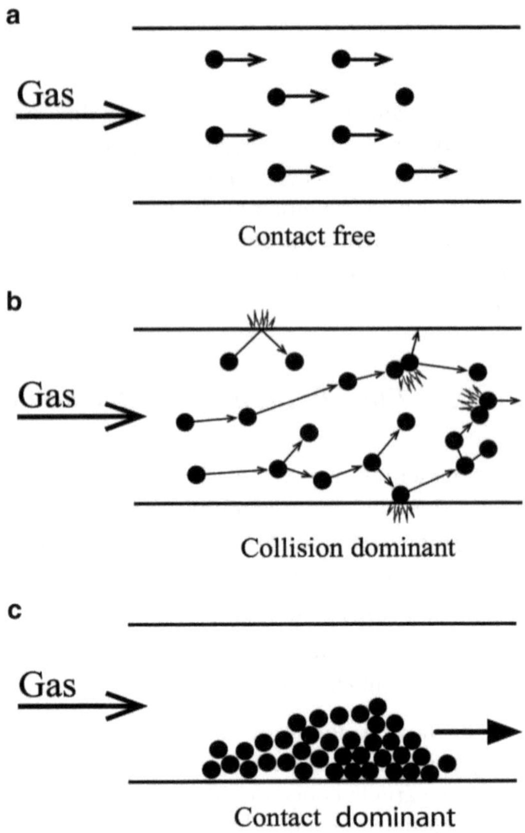

Fig. 10.2 Three types of particle interactions in gas–solid two-phase flow

10.2.3.3 Solid–Gas Two-Phase Jet

A gas jet loaded with solid particles is called a solid–gas two-phase jet (see Fig. 10.5). In materials engineering, this type of jet can be typically seen in a gas atomizer. Many fine metal particles can be produced by this device. This jet can be divided into three regimes with respect to the mixing ratio as follows [13]:

- *Rarefied concentration regime*: The mixing ratio is less than about 0.1 in this regime. The mean velocity and turbulence components of the flow in the gas phase are not significantly affected by the solid particles in the jet.
- *Low concentration regime*: The mixing ratio ranges from about 0.1 to about 5. Solid particles affect the dispersion behavior of the gas phase. Collision among solid particles can be ignored.
- *High concentration regime*: The mixing ratio is higher than about 5. Collision of solid particles cannot be ignored. Solid particles form many clusters in a pipe under specific conditions.

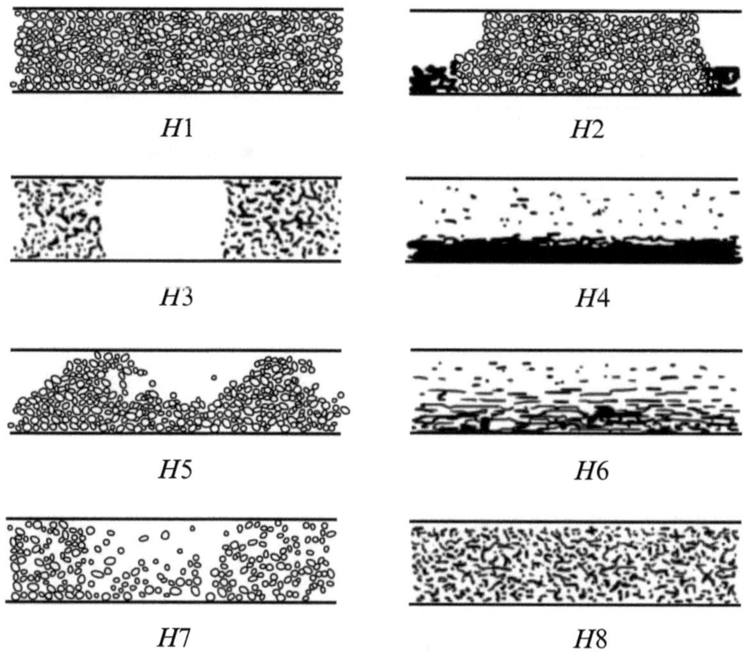

Fig. 10.3 Flow patterns of gas–solid two-phase flow in a horizontal pipe

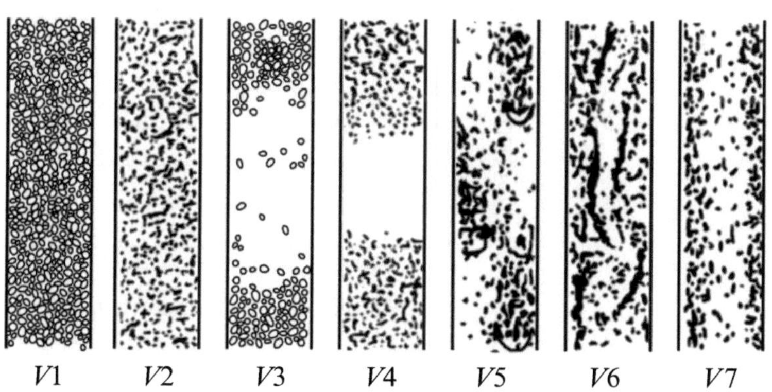

Fig. 10.4 Flow patterns of gas–solid two-phase flow in a vertical pipe

10.2.3.4 Fluidized Bed

The flow pattern in a fluidized bed can be classified as a function of the superficial velocity of gas and the weight-mean particle diameter as follows [14] (see Fig. 10.6):

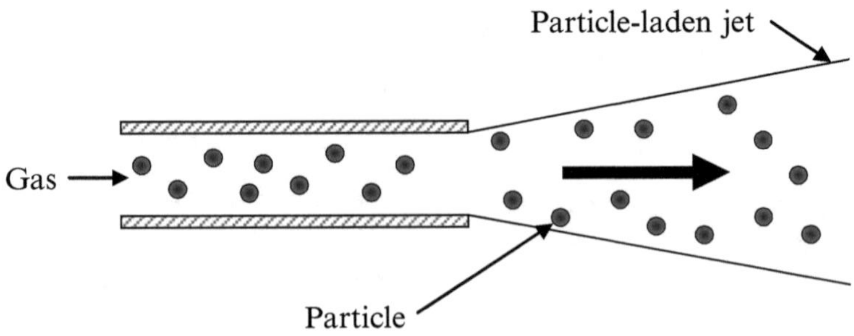

Fig. 10.5 Particle-laden gas jet

Fig. 10.6 Flow patterns of fluidized bed

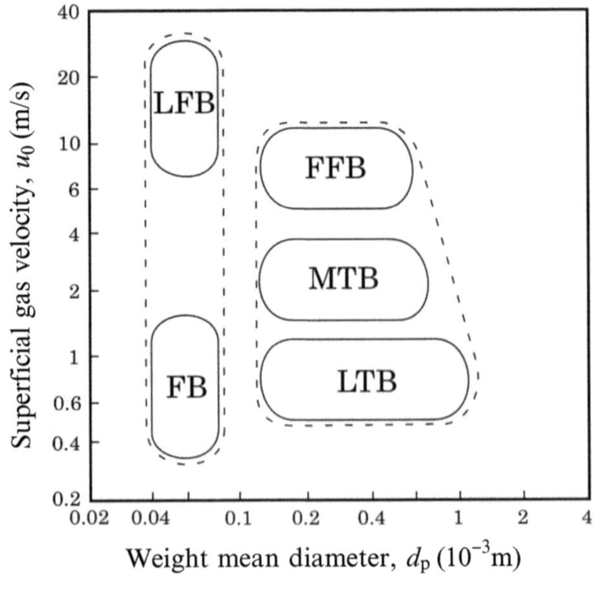

- Coarse-particle fluidized bed or teeter bed (TB): low-speed TB (LTB) and middle-speed TB (MTB)
- Fine-particle fluidized bed or fluid bed (FB)
- Fast fluidized bed (FFB)
- Lean phase fluidized bed (LFB)

10.2.4 Solid–Liquid Two-Phase Flow

10.2.4.1 Solid–Liquid Two-Phase Flow in a Vertical Pipe

The flow patterns of gas–liquid two-phase flow in a vertical pipe were systematically investigated by Sakaguchi et al. [15]. The study indicated that the flow pattern depends strongly on the diameter ratio, d_p/D, where d_p is the particle diameter and D is the pipe diameter.

10.2.4.2 Solid–Liquid Two-Phase Flow in a Horizontal Pipe

A variety of flow types are common in a horizontal pipe including uniform flow, pseudo-uniform flow, and saltation flow [9–16]. Although slurry is regarded as a solid–liquid two-phase flow, the description of such a system will be referred to a subsequent section.

10.2.5 Liquid–Liquid Two-Component One-Phase Flow

Consider the process of pouring molten steel from the ladle to the tundish in a continuous casting operation. The molten slag initially covering the steel in the ladle preferably issues before the steel is completely led to the tundish, as shown in Fig. 10.7. The slag issues while swirling the ladle axis [17]. This phenomenon is called selective withdrawal of liquid from a tank. A similar phenomenon can be observed in a tank filled with two immiscible liquids.

10.2.6 Gas–Liquid–Liquid Two-Phase Flow

Petroleum is sometimes sucked up from the deep underground level through the gas-lift effect [18]. The buoyancy forces acting on air bubbles injected into a vertical long pipe pull petroleum upwards together with the bubbles. In such pipelines seawater is usually mixed in the pipe and, hence, the flow in the pipe becomes a gas–liquid–liquid two-phase flow. Consequently, investigation of the dynamic behavior of gas–liquid–liquid two-phase flows is of practical relevance.

10.2.7 Solid–Liquid Two-Phase Flow (Slurry)

The removal of contaminated sediment from a pond, a lake, or a coastal area by making use of pipeline system is a major problem in environmental engineering.

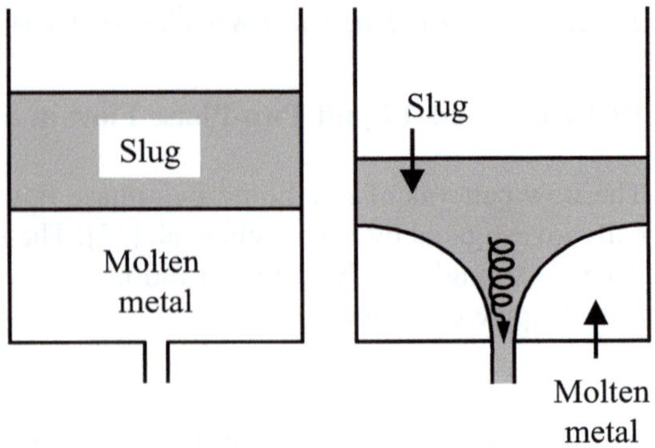

Fig. 10.7 Liquid–liquid two-component one-phase flow

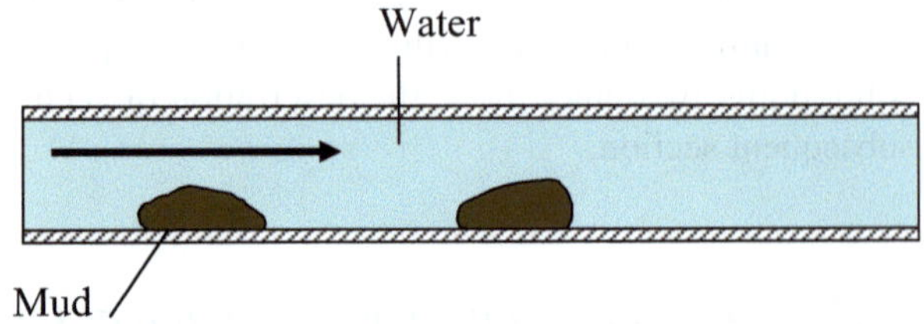

Fig. 10.8 Solid–liquid two-phase flow (slurry)

A solid–liquid two-phase flow called slurry is observed in such a pipeline (see Fig. 10.8). Detailed information on this kind of flow should be referred to Hetsroni [19].

10.2.8 Gas–Liquid–Solid Three-Phase Flow

Figure 10.9 shows a method of sediment removal at the bottom of the sea. The flow in the vertical pipe is an air–seawater–sand three-phase flow.

10.3 Experimental Techniques

10.3.1 Shape, Size, Number, and Moving Velocity of a Solid Particle

The shape and size of a solid particle can be measured by removing it from the flow field. The number of solid particles can be similarly determined precisely. This method, called the in vitro method, however is not necessarily easy and sometimes disturbs the flow field significantly. The measurement method proposed originally

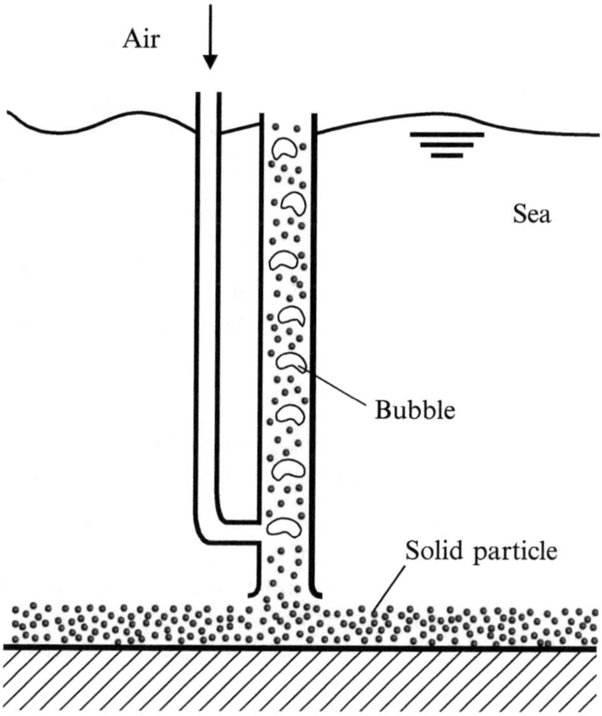

Fig. 10.9 Gas–liquid–solid three-phase flow

in powder and fine-particle engineering is applicable to the present case [20–22]. Considering these circumstances, many in situ techniques have been developed as briefly described below.

10.3.1.1 In Vitro (Indirect) Measurement Method [9, 20, 22]

1. *Screen*: Screens are used for discriminating solid particles of different diameters (see Fig. 10.10). The accuracy of the measurement increases with increased number of screens.
2. *Image processing method*: The shape, size, and number of particles are determined by using a microscope. These parameters are now easily measured due to recent rapid development of image processing algorithms. Even highly overlapped particles can be analyzed with sufficient accuracy [7].
3. *Sedimentation method:* One of the particles removed from the flow field is dropped onto the surface of an adequately chosen solvent bath. Provided that the particle is sufficiently small that Stokes' law of hydrodynamic drag holds, its falling velocity is proportional to the diameter. The diameter thus determined is called the Stokes diameter. The velocity of a particle falling together with many other particles also can be measured with this technique [23].

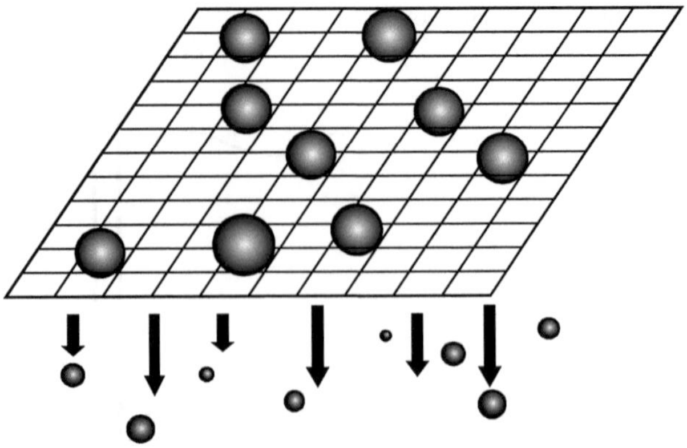

Fig. 10.10 Screen discriminating the size of solid particles

4. *Laser-based optical method:* The following methods have been developed. The measurement accuracy should however be verified through calibration [23].
   (a) *Mie scattering method*: The measurement principle of this technique is based on the elastic scattering characteristics of a particle moving in a parallel laser beam. It employs the relationship between the intensity of scattered light and the particle diameter.
   (b) *Passing light intensity method*: The intensity of a parallel laser beam decreases in the course of passing through a flow field containing solid particles due to scattering and adsorption. This phenomenon enables measurement of the concentration of particles. For example, the mass concentration of soot in a flame can be obtained using the technique.
5. *Diffraction method*: The Fraunhofer diffraction has been applied to the measurement of a representative diameter of solid particles dispersed in a flow filed [24].
6. *Holographic method*: An interaction technique of using pulsed laser has been proposed for measuring the particle diameter and the 3-D position [24].
7. *Correlation method*: The scattering characteristics of laser beam resulting from the Brownian motion of fine particles is employed to measure the particle diameter [24].

10.3.2 Coulter Counter

The Coulter counter has widely been used to measure the representative diameter and the number of particles contained in a liquid simultaneously (see Fig. 10.11). When solid particles pass through a small diameter pipe immersed in an electrically conducting liquid, the resistance between the inlet and outlet of the pipe varies depending on the number and size of particles. The number and size of alumina (Al_2O_3) in molten aluminum can be measured by this method. The development of this type of counter applicable to the steelmaking industry has been extensively investigated. The method can readily detect particles with diameter ranging from about 2 to 40 μm [9, 22].

Fig. 10.11 Coulter counter

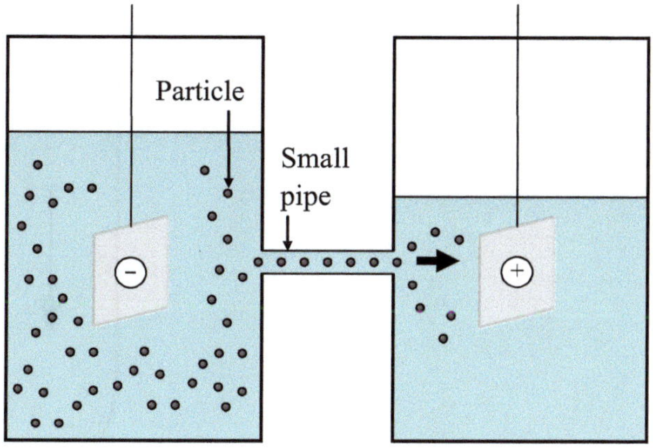

10.3.3 Sheath Flow Method

The number of red cells in human blood can be counted by the sheath flow method. A small pipe is placed upstream of the entrance of a rectangular nozzle [25], as shown in Fig. 10.12. Blood is supplied through the pipe into the nozzle. By adjusting the flow rates of the sheath flow and blood flow, the red cells are introduced into the test section one by one. The laser signal scattered from the cells is analyzed to give the number of red cells. The signal thus detected is dependent on the nature of the particle, enabling the white cell to be discriminated from the red cell. This method also is applicable to the measurement of solid particles.

10.3.4 Ultrasonic Sensor

Ultrasonic sensor has been developed to detect nonmetallic inclusions in molten steel. The magnitude of ultrasonic wave attenuates in the course of passing through a liquid containing many fine solid particles, due to the viscous dissipation caused by the oscillatory motion of the particles. The number and mean size of the particles can be detected on the basis of this principle [26, 27].

10.3.5 In Situ (Direct) Measurement Technique

10.3.5.1 Still Camera

As in the case of gas–liquid two-phase flow measurements, some of the characteristics of dispersed solid particles can be measured by means of still and high-speed video cameras.

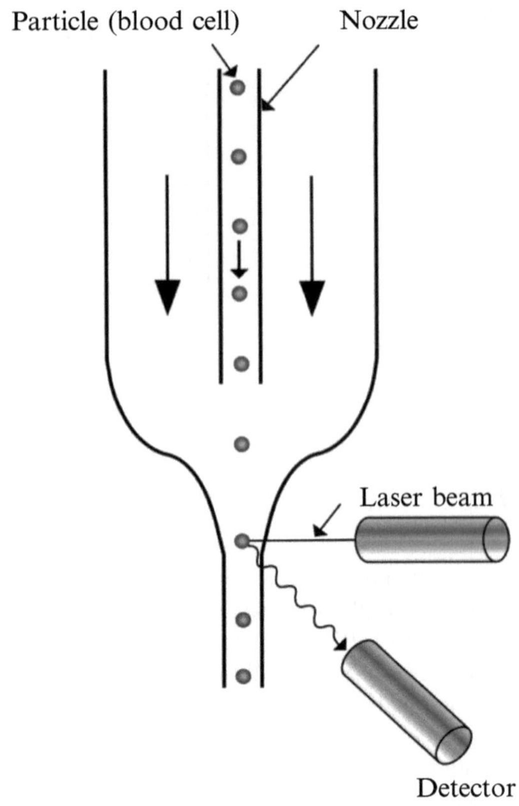

Fig. 10.12 Sheath flow method

10.3.5.2 Laser Doppler Velocimetry (LDV)

LDV is one of the velocimeters capable of providing the local velocity in transparent single-phase gas and liquid flows. The velocity of the fluid is not directly measured but the velocity of a fine seeding particle. The seeding particle is typically spherical and its diameter is about 1 μm. Accordingly, the velocity of fluid flow is obtained only when the seeding particle is small enough to follow the fluid motion. This implies that the velocity of a solid particle can be accurately measured by an LDV as long as the volume of particle is less than the measurement volume, i.e., the intersection zone of the laser beams. In addition, the number of solid particles can be obtained through slight modification of the original data processing program.

A phase Doppler velocimetry has been used to measure the size of particles in addition to the velocity and number of particles [28]. This velocimetry relies on the fact that the phase difference between the Doppler signals from a particle detected at two different positions is proportional to the particle diameter. It should be noted that even the shape of a particle can be measured by the shadow Doppler velocimetry.

Kanda et al. [29] investigated the turbulence characteristics in a solid–gas two-phase wall jet by a three-beam LDV. The technique simultaneously measured the velocities of gas.

10.3.5.3 Particle Image Velocimetry (PIV)

The velocities of many particles can be obtained at once with PIV [30]. Most of the currently used PIV have a function of counting the number of particles. If the color of a particle is changed depending on the material of the particle, the velocity of each material can be obtained by using adequate filter. A particle tracking velocimetry (PTV) technique is suitable for obtaining the particle velocity when the number of particles is not high.

Song et al. [31] developed a novel method for measuring the number and velocity of solid particles using PIV and applied it to the measurement of the number of solid particles falling in the atmosphere and in a water column. This method is useful for counting the number of coke and iron ore supplied into the blast furnace.

10.3.5.4 Optical Method

The optical sensor works based on the photoelectrical effect. When a solid particle crosses a laser beam of small diameter, the voltage of the receiving optics varies drastically. The number of solid particles crossing the beam per unit time can be detected by counting the drastic change in the output voltage. The measurement accuracy does not depend on the physical properties of the particle. However, this technique is not suitable for a flow field containing many particles. The measurement accuracy decreases when more than two particles cross the beam simultaneously.

Kurobe et al. [32] carried out water model experiments on the dynamic behavior of top and bottom dross in the hot-dip plating bath. Figure 10.13 is a schematic sketch of the experimental apparatus. The diameter of the laser beam is 1 mm. Figure 10.14 shows an example of the output voltage of the receiving optics. This signal resembles that of an electroresistivity probe and was processed using the data processing system developed originally for electroresistivity measurement [7].

10.3.5.5 Electrical Capacitance Probe

This probe relies on the fact that the electrical capacitance between two electrode needles immersed in a solid–gas two-phase flow varies when a solid particle passes between them [33, 34].

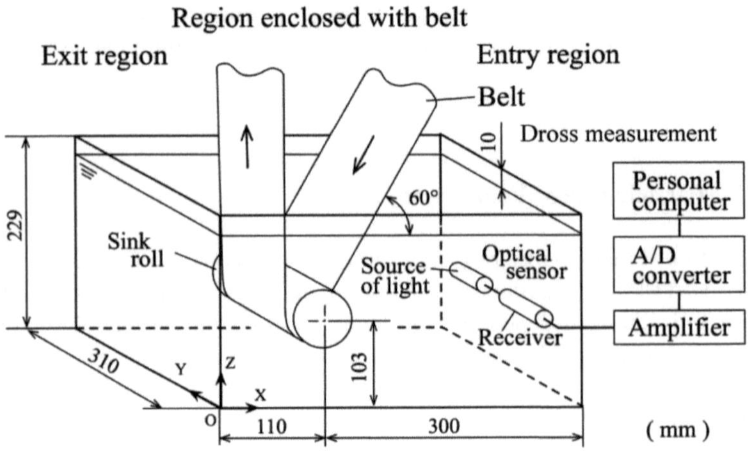

Fig. 10.13 Water model for hot-dip plating bath

$$f_p = N / t_M \quad \text{(Hz)}$$
$$a_p = (\sum t_{pi} / t_M) \times 100 \quad (\%)$$

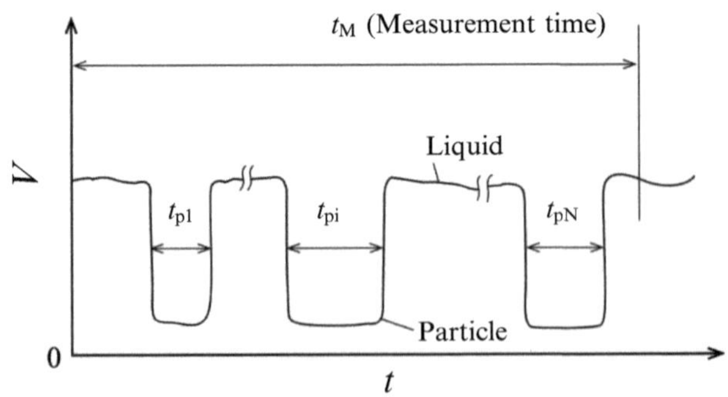

Fig. 10.14 Output signal of optical sensor

10.3.5.6 Optical Fiber Method

The diameter of a particle in solid–gas and solid–liquid two-phase flows can be measured by analyzing the light reflected from the particle [14]. The use of more than two optical fibers enables measurement of the shape of a bubble in a three-dimensional fluidized bed [34, 35].

10.3.5.7 Electrical Capacitance Tomography (ETC)

The electrical capacitance tomography detects the electrical capacitance of a particle using many electrodes and reconstructs its image based on the electrical capacitance data thus detected [16]. The number of electrodes is typically about 6–12. The technique has the capability for real-time measurement.

10.3.5.8 Nuclear Magnetic Resonance Imaging (NMRI)

This method is mostly used for the measurement of solid–liquid two-phase flows, particularly the liquid velocity [16].

10.3.5.9 Visualization Using Ultrasonic Wave

Watanabe et al. [36] visualized water flow containing fine solid particles by the ultrasonic CT method to obtain the velocity of each particle. The particle concentration can be measured by focusing on the damping of the intensity of ultrasonic waves [37].

10.3.5.10 X-ray Visualization Method

Saito et al. [38] observed the clustering of particles in a fluidized bed using X-ray and a high-speed video camera.

10.3.5.11 Hot-Wire Method

The constant-current hot-wire anemometer technique has been applied to detect a bubble in solid–gas two-phase flows [34].

10.3.5.12 Measurement of Mixing Ratio and Concentration

Interested readers should refer to other specialized texts [39–41].

10.4 Concluding Remarks

Many devices have been developed for the measurement of solid–gas and solid–liquid two-phase flows. Unfortunately, most of these devices cannot be applicable to flows in many materials processing operations, mostly in the steelmaking industry due to the high fluid temperature. Further investigations are desirable to develop novel measurement techniques applicable to multiphase flows at very high temperature.

References

1. ISIJ (1976) Modeling theory and scale-up in steelmaking, ISIJ, Tokyo
2. Sato M (2000) Analysis of powder behavior in blast furnaces. J Jpn Soc Multiphase Flow 14–4:426–433
3. JSPS (1994) 19th Committee: recent investigations on non-metallic inclusions in molten steel. JSPS, Tokyo
4. Taniguchi S, Kikuchi A (1992) Mechanisms of collision and coagulation between fine particles in fluid. Tetsu Hagane 78–4:527–535
5. Szekely J (1979) Fluid flow phenomena in metals processing. Academic, New York
6. Ilegbusi OJ, Iguchi M, Wahnsiedler W (1999) Mathematical and physical modeling of materials processing operations. Chapman-Hall/CRC, Voca Raton
7. Iguchi M (2001) Gas-liquid two-phase flow. Ferrum 6–10:793–799
8. Morikawa Y (1979) Fluid-solid two-phase flow, pneumatic transportation and fluid transportation. Nikkann-Kogyo Shinnbun Co., Ltd
9. Chemical Engineering Society of Japan (1999) Handbook of chemical engineering, 6th edn. Chemical Engineering Society of Japan p 231
10. Masuyama T (2000) Measurement of solid concentration in solids–liquids two-phase flow in pipes. J Jpn Soc Multiphase Flow 14–3:263–269
11. Tomita Y, Harada S (2000) Flow pattern and visualization of gas–solid two-phase flow. J Jpn Soc Multiphase Flow 14–2:158–165
12. Tsuji Y (1984) Fundamentals of pneumatic transportation. Yokendo Book Co., Ltd., Tokyo, p 57
13. Fluid Dynamics (1991) Dynamics of multiphase flows. Asakura Book Co., Ltd., Tokyo, p 179
14. Chiba T, Yoshida K (1996) Fluidized bed. Asakura Book Co., Ltd., Tokyo, p 1
15. Sakaguchi T, Hosokawa S, Ohe S, Tomiyama A (1993) Study on flow patterns in large particle-water two-phase flow. Proc Multiphase Flow Symp 93:123
16. Kasagi N et al(1997) Handbook of fluid flow experiments. Asakura Book Co., Ltd., Tokyo, p 600
17. Hammerschmidt P, Tacke KH, Popper H, Weber L (1984) Dubke M and schwerdtfeger. Ironmaking Steelmaking 11–6:332–339
18. Hetsroni G (1982) Handbook of multiphase systems. Hemisphere, Washington, pp 7–30
19. Hetsroni G (1982) Handbook of multiphase systems. Hemisphere, Washington, pp 9–98
20. Jinbo G et al(1989) Handbook of fine particles. Asakura Book Co. Ltd, Tokyo
21. Kataoka S et al(1986) Sensor handbook. Baifukan Book Co., Ltd., p 449, 477
22. Powder Engineering (1998) Handbook of powder engineering, 2nd edn. Nikkan Kogyo Shinbun Co., Ltd., Tokyo, p 3
23. Umekage T, Kajimoto K, Yu S (1996) Measurement of drag force acting on spherical particle in multiparticle system. Proc Multiphase Flow Symp 96:235–238
24. Editorial Board for Publishing Handbook of Laser Measurement (1993) Handbook of Laser Measurement. Maruzen Book Co., Ltd., Tokyo, p 229
25. Fujimoto K (1999) Review on sheath flow. Sysmex J. In 9–1 Kobe, p 31
26. Strout TA (1991) Attenuation of sound in high-concentration suspensions: development and application of an oscillating cell model. Ph.D. Thesis, University of Maine, Prono, ME
27. Matsuoka T, Kitano H, Yasuda K, Koda S, Nomura H (1997) Determining the size of colloidal particles using ultrasonic absorption measurements. Jpn J Apl Phys 36:2972–2975

28. Hishida K (2000) Laser technique for gas-solid two-phase flow: measurements of velocity, size and concentration. J Jpn Soc Multiphase Flow 14–2:150–157
29. Kanda M, Sato Y, Hishida K, Maeda M (1995) Particle response and turbulence modification in gas-solid wall jet. Proc Multiphase Flow Symp 95:116
30. Song X, Yamamoto F, Iguchi M (2001) Image processing (Particle image velocimetry). Ferrum 6–4:250–258
31. Song X, Yamamoto F, Iguchi M, Kato H, Murai Y, Ishii K, Sasaki Y (1999) Particle imaging velocimetry measurement of particle number rate and velocity for falling particles. ISIJ Int 39:471–476
32. Kurobe J, Iguchi M (2001) Cold model experiment on dynamic behavior of dross in hot dip plating bath. Sci Technol Adv Mater 1:251–259
33. Saito T (2000) Measurements of dispersed particle velocity in solid-liquid two-phase flows. J Jpn Soc Multiphase Flow 14–3:302–309
34. Chiba T, Yoshida K (1996) Fluidized bed. Asakura Book Co., Ltd., Tokyo, p 229
35. Hatano H, Takeuchi H (2000) Visualization of fluidized bed behavior. J Jpn Soc Multiphase Flow 14–4:418–425
36. Watanabe K, Che F, Kokubo K (1997) Flow visualization of water/fine solid particle suspensions by ultrasonic. Proc Multiphase Flow Symp 97:11–12
37. Tsuji Y (1984) Fundamentals of pneumatic transportation. Yokendo Book Co., Ltd., Tokyo, 190.
38. Saito K, Ueda Y, Kawata Y, Watanabe S (1996) Visualization with X-ray and numerical simulation of solid–gas multiphase flow (Comparison between experiment and simulation of circulating fluidized bed). Proc Multiphase Flow Symp 96:165–168
39. IIdani K et al(1970) Powder engineering handbook, 5th edn. Asakura Book Co., Ltd., Tokyo, p 662
40. Powder Engineering (1998) Handbook of powder engineering, 2nd edn. Nikkan Kogyo Shinbun Co., Ltd., p 461
41. Watanabe K (2000) Detecting flow-rate of powder and bulk solids in gas–solids two phase flow. J Jpn Soc Multiphase Flow 14–2:144–149

Chapter 11
Multiphase Flow in Microscale Systems

Abstract The surface and interfacial tension forces play an essential role in micro reactors and channels. Wall wettability is also one of key parameters affecting the flow in these flow systems. Detailed discussion is given on the flows in microscale systems in this chapter.

Keywords Capillary number • Dynamic contact angle • Equilibrium contact angle • Fuel cell • MEMS • Micro channel • Micro reactor • Weber number • Wettability

11.1 Introduction

The scale of the currently used micro reactors ranges from the order of micrometer (μm) to millimeter (mm) [1–6]. The channels of the reactors also have the same order of magnitude and therefore are collectively called micro channels. In the conventional channels with size ranging from centimeter (cm) to meter (m), body forces such as the gravitational, centrifugal, and buoyancy largely govern the fluid dynamics. On the other hand, the viscous, interfacial, and surface forces begin to play important role in micro channels. In other words, fluid flow in micro channels is primarily described by the Reynolds number, *Re*, and the Weber number, *We*, defined as

$$Re = \frac{VL}{\nu_f} \tag{11.1}$$

$$We = \frac{\rho V^2 L}{\sigma} \tag{11.2}$$

where V is the representative velocity, L is the representative length, ν_{f} is the kinematic viscosity of the fluid, ρ is the density of the fluid, and σ is the surface tension.

M. Iguchi and O.J. Ilegbusi, *Basic Transport Phenomena in Materials Engineering*,
DOI 10.1007/978-4-431-54020-5_11,

11.2 Basic Characteristics of Gas–Liquid Two-Phase Flow in Micro Channel

11.2.1 Flow Patterns in Horizontal Circular Micro Channels

Previous investigations on micro channels have considered mostly horizontal channels due to their practical relevance to micro reactor technology. Some of these investigations are briefly reviewed below.

Fukano et al. [7] carried out experimental investigations on flow pattern and pressure drop in small-scale circular pipes. The pipe diameter ranged from 1.0 to 4.9 mm. The map of the flow pattern agreed with that proposed by Barnea et al. [8].

Chung and Kawaji [9] used circular pipes ranging in size from 50 to 530 μm in diameter to investigate the flow pattern, void fraction, and pressure drop. The results indicate bubbly, slug, and annular flows. In particular, slug flow was evident mainly in the smallest (50 μm and 100 μm) pipes. Serizawa et al. [10] used circular pipes of 20 to 100 μm in diameter to investigate the flow pattern and void fraction, observing also similar bubbly, slug, and annular flows. It was concluded that the slug flow would likely be affected by the contamination of the liquid and the wettability of the pipe wall. Ide et al. [11] and Kawahara et al. [12] groups investigated the relationship between the flow pattern and the methods used to supply the gas and liquid at the pipe entrance.

11.2.2 Flow Pattern in Horizontal Rectangular Micro Channels

Experimental investigations on the flow pattern in horizontal rectangular micro channels have been carried out by Sadatomi et al. [13], Ide and Fukano [14], and Kumagai and Iguchi [15].

11.2.3 Flow Pattern in Horizontal Channels with Arbitrary Cross Section

In a review study, Akbar et al. [16] classified the entire flow pattern map into four regimes: surface tension-dominant region (bubbly, plug, and slug flows); inertia-dominant zone 1 (annular and wavy–annular flows); inertia-dominant zone 2 (dispersed or froth flow); and transition zone.

The first three regimes can be described as functions of the gas Weber number, We_{GS}, and the liquid Weber number, We_{LS}, as follows:

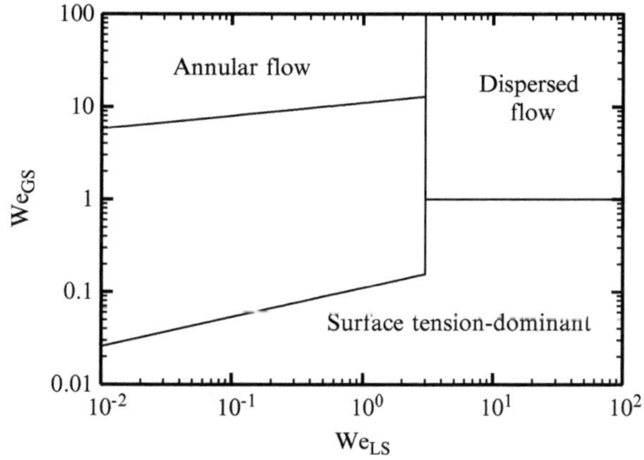

Fig. 11.1 Flow pattern map for gas–liquid two-phase flow in microchannels

11.2.3.1 Surface Tension-Dominant Zone (see Fig. 11.1)

$$We_{GS} \leq 0.11 We_{LS}^{0.315} \tag{11.3}$$

$$We_{LS} \leq 3.0 \tag{11.4}$$

and

$$We_{GS} \leq 1.0 \tag{11.5}$$

$$We_{LS} > 3.0 \tag{11.6}$$

11.2.3.2 Annular Flow Zone (Inertia-Dominant Zone 1)

$$We_{GS} \geq 11.0 We_{LS}^{0.14} \tag{11.7}$$

$$We_{LS} \leq 3.0 \tag{11.8}$$

11.2.3.3 Dispersed Flow Zone (Inertia-Dominant Zone 2)

$$We_{GS} > 1.0 \tag{11.9}$$

$$We_{LS} > 3.0 \tag{11.10}$$

where

$$We_{GS} = \frac{\rho_G V_{GS}^2 D_h}{\sigma} \quad (11.11)$$

$$We_{LS} = \frac{\rho_L V_{LS}^2 D_h}{\sigma} \quad (11.12)$$

$$V_{GS} = \frac{Q_G}{A} \quad (11.13)$$

$$V_{LS} = \frac{Q_L}{A} \quad (11.14)$$

$$D_h = \frac{4A}{P} \quad (11.15)$$

Here ρ_G is the density of the gas, V_{GS} is the superficial velocity of the gas, D_h is the hydraulic diameter, σ is the surface tension, ρ_L is the density of the liquid, V_{LS} is the superficial velocity of the liquid, Q_G is the gas flow rate, Q_L is the liquid flow rate, A is the cross-sectional area of the pipe, and P is the peripheral length of the pipe.

11.2.4 Pressure Drop

Mishima and Hibiki [17] proposed a modified Lockhart–Martinelli method to predict the pressure drop in micro channels.

11.3 Effect of Pipe Wettability on Gas–Liquid Flow in Micro Channel

11.3.1 Equilibrium Contact Angle, θ_c

The wettability should be discussed from two points of view: static and dynamic. The static wettability is quantitatively described by the equilibrium contact angle, θ_c. Consider a droplet placed on a horizontal flat plate, as shown in Fig. 11.2. If the plate is wetted by the droplet, the droplet spreads on the plate. On the other hand, if the plate is poorly wetted, the droplet tends to be spherical in shape. When the equilibrium contact angle, θ_c, is less than 90° (0° $\leq \theta_c <$ 90°), the wettability is

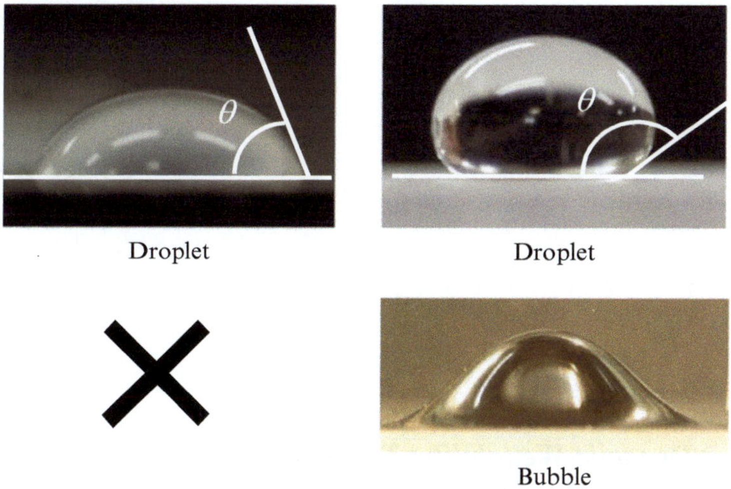

Fig. 11.2 Droplet and bubble on horizontal flat plate

considered good (good wettability) and poor for $90° \leq \theta_c \leq 180°$ (poor wettability). The equilibrium contact angle of molten steel, θ_c, ranges from about 130° to 150°.

When a poorly wetted flat plate is submerged horizontally in a water bath, as shown in Fig. 11.2, air can remain on the upper surface of the plate against the buoyancy force acting on it [18]. As the equilibrium contact angle increases, the air bubble begins to spread on the plate. Meanwhile, when the plate is wetted by water, air cannot stay on the upper surface anymore. The implication is that gas can completely cover the surface of a plate when $\theta_c = 180°$.

The wall of steelmaking reactors may be partly covered with injected gas such as argon and oxygen even when $\theta_c \approx 130° \sim 150°$. The same situation is realized when a mixture of gas and liquid is introduced into a poorly wetted pipe. The gas tends to preferably attach to the pipe wall and, thus, the flow pattern and pressure loss would differ from those of gas–liquid two-phase flows in a wetted pipe. This presumption was confirmed by Iguchi et al. [19, 20] for air–water two-phase flows in a poorly wetted vertical pipe. The most interesting phenomenon was the appearance of reversed annular flow in which air flows along the pipe wall, while water flows in the central part of the pipe.

Concerning the pressure loss, the wettability effect seems to be dependent on the size of the pipe and the flow pattern. Systematic investigations are desirable in order to fully understand this effect. It should be noted that although the wettability of a plate can be changed by coating the surface with water repellent, the life of the repellent is not necessarily long. For example, if the initial equilibrium contact angle of a repellent-coated wall is $\theta_c = 147°$, the angle progressively decreases with time due to contamination. Therefore care must be taken in carrying out such experiments [21].

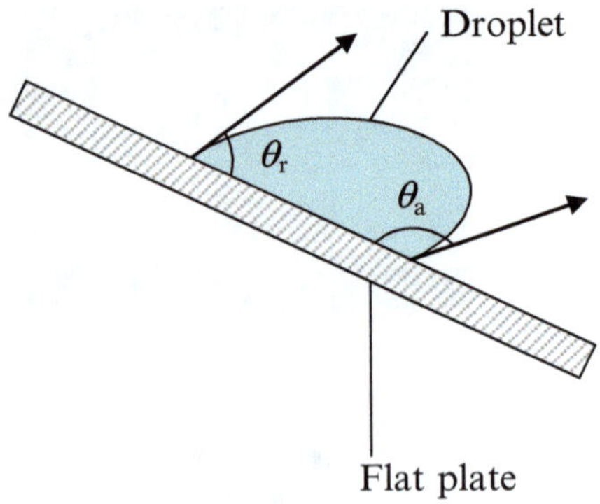

Fig. 11.3 Advancing and receding contact angles

11.3.2 Dynamic Wettability: The Advancing and Receding Contact Angles

In a strict sense the behaviors of gas–liquid two-phase flows in a poorly wetted pipe should be discussed on the basis of dynamic wettability. Specifically, it is necessary to introduce the concept of advancing and receding contact angles, θ_a and θ_r (see Fig. 11.3). At present, however, the measurement of θ_a and θ_r is rather difficult and, hence, there is yet no quantitative discussion on the effects of dynamic wettability on the basic characteristics of gas–liquid two-phase flows in a poorly wetted pipe.

11.4 Gas–Liquid Flow Through Abrupt Expansion in Micro Channel

Micro channels often contain a variety of pipe elements, typical of which are abrupt expansion, abrupt contraction, and orifice. The following discussion is concerned with air–water two-phase flows passing through an abrupt expansion and an orifice.

Oke et al. [22] carried out experiments using horizontal rectangular micro channels with constant width of 50 mm and height ranging from 190 μm to 2 mm (see Fig. 11.4). The main objective of the study was to understand the effect of an abrupt expansion on the flow pattern in a gas–liquid two-phase flow in millimeter-scale rectangular channels. The flow patterns were categorized mainly into four types:

1. The flow pattern is not influenced by the abrupt expansion. That is, the flow patterns upstream and downstream of the expansion are the same. This situation was often observed in the small expansion ratios.
2. A recirculation zone appears just after the abrupt expansion. This flow pattern was observed for high water flow rate and large expansion ratio.

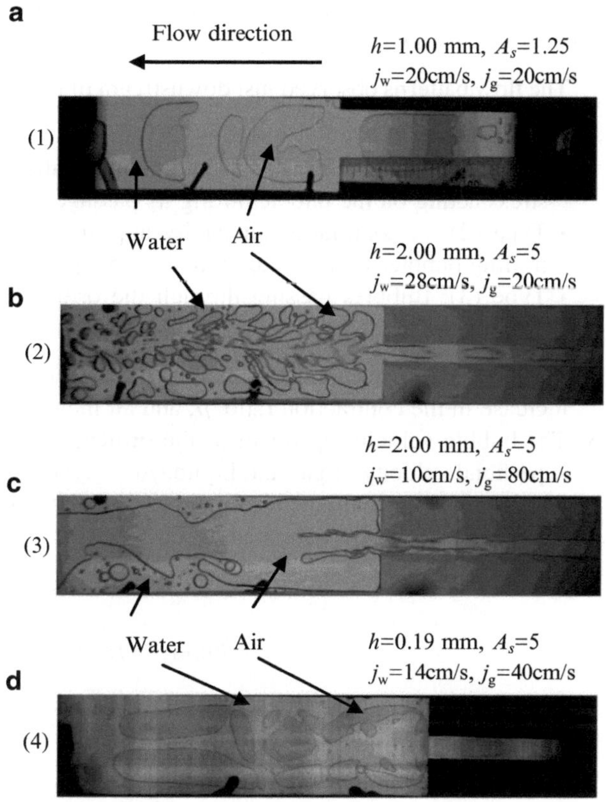

Fig. 11.4 Typical flow patterns of gas–liquid two-phase flow through the abrupt expansion

3. An air cavity is formed just downstream of the abrupt expansion and a water jet penetrates into it. This pattern was observed in high air flow rate, low water flow rate, and large expansion ratio.
4. Each bubble splits in the lateral direction just after the abrupt expansion.

A recirculation zone was observed downstream of the abrupt expansion for $h = 0.50$–2.00 mm at high expansion ratios, similar to the pattern observed in the conventional size pipes. By contrast, an interesting phenomenon was observed at $h = 0.19$ mm. In this flow pattern, each bubble splits in the lateral direction without formation of the recirculation zone. The effects of volumetric forces disappear and the surface force becomes dominant.

11.5 Gas–Liquid Flow Through Orifice in Micro Channel

Oke et al. [23] experimentally investigated the effects of an orifice on the flow pattern and bubble velocity in gas–liquid two-phase flows in millimeter-scale rectangular channels. The bubbly, slug, and annular flows were observed in the test section upstream of the orifice. The major findings of this study are summarized below.

1. The flow patterns observed just downstream of the orifice could be classified into the following three types:
   - Type (1): A bubble breaks up into many smaller bubbles due to strong shear stress acting on the bubble during its passage through the orifice.
   - Type (2): A continuous air cavity appears downstream of the orifice. The trailing bubbles join the cavity and water spreads in the air cavity.
   - Type (3): Bubbles passing through the orifice are squeezed and elongated. There is no breakup of bubble never into smaller bubbles.
2. The surface force becomes dominant with a decrease in the channel height, h, an increase in the contraction ratio, β, and an increase in the orifice thickness, L_o.
3. The bubble velocity upstream of the orifice, $u_{B,up}$, is nearly equal to the total superficial velocity of gas and liquid,($j_T = j_G + j_L$). The maximum bubble velocity, $u_{B,max}$, is approximated by Eq. (11.16) in the low j_T/β regime. In the high j_T/β regime, $u_{B,max}$ is approximated by Eq. (11.17):

$$u_{B,max} = j_T/\beta \tag{11.16}$$

$$u_{B,max} = 0.5 j_T/\beta \tag{11.17}$$

11.6 Examples of Microscale Devices

Micro reactors are widely used for producing materials [24] and for diagnostic applications [25, 26]. The following are examples of typical microscale devices.

11.6.1 Microfluidics

According Stone et al. [5], microfluidics refers to devices and methods for controlling and manipulating fluid flows with less than a millimeter. Recent developments in microfluidics are reviewed by Stone et al. [5] and McCalla et al. [26].

11.6.2 Fuel Cells

Micro channels are commonly seen in fuel cells. On the cathode of the fuel cells oxygen and hydrogen react to generate water. The drainage of such water is of practical significance for enhancing the efficiency of the fuel cell. It is quite difficult to drain only water through micro channels because air and the remaining oxygen and hydrogen are introduced together into the channels. The development of an effective drainage method is one of the key technologies necessary for stable operations of the fuel cells [27, 28].

11.6.3 Lab-on-a-Chip

Lab-on-a-chip devices are commonly used in biomedical engineering [5, 29].

11.6.4 Cooling System for Electronic Devices Such as Personal Computers

Air cooling method has conventionally been used for cooling electric and electronic devices such as computer chips. Recently, there has been significant interest in the use of micro channels and water in place of the current air cooling method. This is because micro channels have very high surface-to-volume ratio and, hence, high heat-transfer rate can be realized by stacking many micro channels.

11.6.5 Microelectromechanical Systems (MEMS)

MEMS have recently been extensively used in fluid measurement and control [30–32].

11.6.6 Generation of Micro- and Nanoparticles

Production of droplets and particles smaller than micrometer in diameter are of relevance to many engineering fields. Examples include production of micro- and nanocapsules, coaxial nanofibers, and hollow nanofibers. These droplets and particles can be generated by using capillary flows [33–35].

11.7 Concluding Remarks

The role of multiphase flows in microscale systems is increasingly important. Flow control in such systems would be a major concern deserving of future investigation. This chapter has presented a brief description of flow phenomena in micro channels mostly from experimental observation. It also provides a list of typical embodiment of microscale devices in industry.

References

1. Fujii T (2008) Development and applications of micro reactor. In: Yoshida Y (ed) CMC, Tokyo, pp 47–60
2. Kusakabe K, Sotowa K (2008) Introduction to micro reactor. Yoneda Pub. Co. Ltd., Tokyo, 43–78
3. Togashi S, Miyake R (2005) Micro reactor technology. NTS Co. Ltd., Tokyo, 41–49
4. Kawai T (2003) Dictionary of nanotechnology. Kogyo-chosakai, Tokyo, 527
5. Stone HA, Stroock AD, Ajdari A (2004) Engineering flows in small devices: microfluidics toward a lab-on-a-chip. Ann Rev Fluid Mech 36:381–411
6. Meyvantsson I, Beebe DJ (2008) Cell culture models in microfluidic systems. Ann Rev Anal Chem 1:423–449
7. Fukano T, Kariyasaki A, Kagawa M (1990) Flow patterns and pressure drop in isothermal gas–liquid concurrent flow in a horizontal capillary tube. Trans Jpn Soc Mech Eng, Ser B 56 (528):174–182
8. Barnea D, Luninski Y, Taitel Y (1983) Flow pattern in horizontal and vertical two-phase flow in small diameter pipes. Can J Chem Engngi 61:617–620
9. Chung PMY, Kawaji M (2004) The effect of channel diameter on adiabatic two-phase flow characteristics in microchannels. Int J Multiphase Flow 30:735–761
10. Serizawa A, Feng Z, Kawahara Z (2002) Two-phase flow in microchannels. Exp Therm Fluid Sci 26:703–714
11. Ide H, Kimura R, Kawaji M (2010) Characteristics of void fraction and velocity of gas–liquid two-phase flow in a microchannel. J Jpn Soc Multiphase Flow 23–5:589–596
12. Kawahara A, Sadatomi M, Nei K, Matsuo H (2009) Investigation of bubble velocity, void fraction an pressure drop for gas–liquid two-phase flow in a circular microchannel. Progress Multiphase Flow Res 4:37–44
13. Sadatomi M, Ali MI, Kawaji M (1992) Effects of gap width and orientation on two-phase flow in a narrow passage between two-float plates. Trans Jpn Soc Mech Eng Ser B 58(546):112–119
14. Ide H, Fukano T (2003) Correlations of the holdup and frictional pressure drop in air-water two-phase flow in a flat capillary rectangular channel. Trans Jpn Soc Mech Eng Ser B 69 (686):67–75
15. Kumagai T, Iguchi M (2007) Wettability effect on flow pattern in horizontal air-water two-phase flow in rectangular small channels. J JSEM 7(1):50–55
16. Akbar MK, Plummer DA, Ghiaasiaan SM (2003) Brief communication on gas–liquid two-phase flow regimes in microchannels. Int J Multiphase Flow 29:855–865
17. Mishima K, Hibiki T (1996) Some characteristics of air–water two-phase flow in small diameter vertical tubes. Int J Multiphase Flow 22:703–712
18. Watanabe T, Iguchi M (2009) Model experiment on the mold powder entrapment by a large argon bubble rising along the immersion nozzle. ISIJ In 49(2):182–188
19. Terauchi Y, Iguchi M, Kosaka H, Yokoya S, Hara S (1999) Wettability effect on the flow pattern of air-water two-phase flows in a vertical pipe. Tetsu Hagane 85(9):645–651

20. Iguchi M, Terauchi Y (2001) Boundaries among bubbly and slug flow regimes in air–water two-phase flows in vertical pipe of poor wettability. Int J Multiphase Flow 27:729–735
21. Inoue T, Iguchi M, Mizuno M (2001) Separation of gas and liquid using wettability difference of T-junction. Jpn J Multiphase Flow 15(2):158–164
22. Oke T, Kumagai T, Iguchi M (2010) Flow patterns of gas–liquid two-phase flow through an abrupt expansion in millimeter-scale rectangular channel. J JSEM 10(Special Issue):32–37
23. Oke T, Kumagai T, Ilegbusi OJ, Iguchi M (2010) Gas-liquid two-phase flow through an orifice in milimeter-scale rectangular channel. Proceding 5th international symposium on advanced science and technology in experimental mechanics, Kyoto, 4–7 November 2010
24. Livak-Dahl E, Sinn I, Burns M (2011) Microfluidic chemical analysis system. Ann Rev Chem Bio Eng 2:325–353
25. Beebe DJ, Mensing GA, Walker GM (2002) Physics and applications of microfluidics in biology. Ann Rev Biomed Eng 4:261–286
26. McCalla SE, Tripathi A (2011) Microfluidic reactors for diagnostics applications. Ann Rev Biomed Eng 13:312–343
27. Young JB (2007) Thermofluid modeling of fuel cells. Ann Rev Fluid Mech 39:193–215
28. Brandon NP, Skinner S, Steele BCH (2003) Recent advances in materials for fuel cells. Ann Rev Mater Res 33:183–213
29. Toner M, Irimia D (2005) Blood-on-a-chip. Ann Rev Biomed Eng 7:77–103
30. Ho CM, Tai YC (1998) Micro-electro-mechanical-systems (MEMS) and fluid flows. Ann Rev Fluid Mech 30:579–612
31. Kasagi N, Suzuki Y, Fukagata K (2009) Microelectromechanical systems-based feedback control of turbulence for skin friction reduction. Ann Rev Fluid Mech 41:231–251
32. Gutierrez-Osuna R, Hierlemann A (2010) Adaptive microsensor systems. Ann Rev Anal Chem 3:255–276
33. Ghosal S (2006) Electrokinetic flow and dispersion in capillary electrophoresis. Ann Rev Fluid Mech 38:309–338
34. Barrero A, Loscertales IG (2007) Micro- and nanoparticles via capillary flows. Ann Rev Fluid Mech 39:89–106
35. Rauscher M, Dietrich A (2008) Wetting phenomena in nanofluidics. Ann Rev Mater Res 38:143–172

Problems and Answers

Part I

Problem I.1 Two parallel flat plates are immersed vertically in water. The clearance between the two plates is b. Calculate the elevation of the liquid due to capillary force. The surface tension, σ, and the density of water are 0.0725 N/m and 997 kg/m^3, respectively. The contact angle, θ_c, is 5°, and $b = 0.5$ mm.

Problem I.2 Calculate the absolute pressure on the bottom of the sea. The depth of the sea, H, is 250 m, the density of seawater, ρ_{sw}, is 1,030 kg/m^3, and the atmospheric pressure, p_0, is 1,013.3 h Pa.

Problem I.3 Mercury is used as the working fluid for measuring the pressure of water by a U-shaped manometer. The water pressure at one entrance of the manometer, p_1, is 300 k Pa, and pressure at the other entrance, p_2, is 230 k Pa. Calculate the mercury column height, H. The densities of the mercury and water are 13,600 kg/m^3 and 1,000 kg/m^3, respectively.

Answer to Problem I.1

Let the width of each plate be W. The following force balance holds:

$$bWp_0 + \rho gbWH = bWp_0 + 2W\sigma\cos\theta_c \tag{PI.1}$$

The height, H, is expressed by

$$H = 2\sigma\cos\theta_c/(\rho gb) \tag{PI.2}$$

$$\begin{aligned} H &= 2 \times 0.0725 \times \cos\ 5°/(997 \times 9.80 \times 0.5 \times 10^{-3}) \\ &= 0.0296\ \text{m} = 29.6\ \text{mm} \end{aligned} \tag{PI.3}$$

M. Iguchi and O.J. Ilegbusi, *Basic Transport Phenomena in Materials Engineering*,
DOI 10.1007/978-4-431-54020-5,

Answer to Problem I.2

$$\begin{aligned} p &= p_0 + \rho_{sw} g H = 1{,}013.3\,\text{hPa} + 1{,}030 \times 9.8 \times 250\,\text{Pa} \\ &= 101.33\,\text{kPa} + 2523.5\,\text{kPa} = 2.62\,\text{MPa} \end{aligned} \quad \text{(PI.4)}$$

Answer to Problem I.3

The mercury column height is calculated from Eqs. (1–12) as follows:

$$\begin{aligned} H &= (p_1 - p_2)/[(\rho_{Hg} - \rho_w)g] \\ &= (300 - 230) \times 10^3/[(13{,}600 - 1{,}000) \times 9.8] = 0.567\,\text{m} \end{aligned} \quad \text{(PI.5)}$$

Problem I.4 Calculate the Reynolds number, *Re*, for oil flow in a circular pipe. The diameter of the pipe is 50 mm, the density of the oil is 920 kg/m^3, the volumetric oil flow rate is 56 L/min, and the dynamic viscosity of the oil is 40 m Pa s.

Problem I.5 Calculate the mean velocity of water flow in a pipe. The Reynolds number is 3,000, the pipe diameter is 10 mm, and the kinematic viscosity of water is 1.01×10^{-6} m^2/s.

Problem I.6 Consider the flow in a pipe. The pipe diameter is 50 mm, the density of fluid is 920 kg/m^3, the flow rate is 0.150 m^3/min, and the kinematic viscosity is 56 m Pa s. The critical Reynolds number is 2,320. Determine if the flow is laminar or turbulent.

Problem I.7 Calculate the volumetric flow rate of oil flowing in a circular pipe of 50 mm diameter. The Reynolds number is 2,320, the density of oil is 920 kg/m^3, and the dynamic viscosity of oil is 50 m Pa s.

Problem I.8 Show that the following relationship is obtained from the parabolic distribution of laminar pipe flow velocity:

$$v_m/v_{cl} = 1/2 \quad \text{(PI.6)}$$

Problem I.9 Show that the following relationship is obtained from the 1/7th power law of turbulent pipe flow velocity:

$$v_m/\bar{v}_{cl} = 49/60 \quad \text{(PI.7)}$$

Problem I.10 Show that the following relationship is obtained from the 1/*n*th power law of turbulent pipe flow velocity:

$$v_m/\bar{v}_{cl} = 2n^2/[(n+1)(2n+1)] \quad \text{(PI.8)}$$

Answer to Problem I.4

The cross-sectional mean velocity, v_m, is given by

$$v_m = 4Q/(\pi D^2) = 4 \times 56 \times 10^{-3} \times (1/60)/[3.14 \times (50 \times 10^{-3})^2] \\ = 0.476\,\mathrm{m/s} \tag{PI.9}$$

The Reynolds number is calculated to give

$$Re = \rho v_m D/\mu = 920 \times 0.476 \times 50 \times 10^{-3}/(40 \times 10^{-3}) = 547 \tag{PI.10}$$

where ρ and μ are the density and dynamic viscosity, respectively.

Answer to Problem I.5

The Reynolds number is expressed by

$$Re = \frac{v_m D}{\nu_f} \tag{PI.11}$$

where ν_f is the kinematic viscosity. The cross-sectional mean velocity, v_m, is calculated from Eq. (pI.11) to give

$$v_m = \nu_f Re/D = 1.01 \times 10^{-6} \times 3{,}000/(10 \times 10^{-3}) = 0.303\,\mathrm{m/s} \tag{PI.12}$$

Answer to Problem I.6

The cross-sectional mean velocity, v_m, is given by

$$v_m = 4Q/(\pi D^2) = 4 \times 0.150 \times (1/60)/[3.14 \times (50 \times 10^{-3})^2] \\ = 1.27\,\mathrm{m/s} \tag{PI.13}$$

The Reynolds number becomes

$$Re = \rho v_m D/\mu = 920 \times 1.27 \times 50 \times 10^{-3}/(56 \times 10^{-3}) = 1{,}043 \tag{PI.14}$$

This Reynolds number is lower than the critical Reynolds number of 2,320, and the flow is laminar.

Answer to Problem I.7

The flow rate, Q, is expressed by

$$Q = \pi D^2 v_m/4 = (\pi D^2/4) \times \nu_f Re/D = \pi D \nu_f Re/4 = \pi D \mu Re/(4\rho) \\ = 3.14 \times 40 \times 10^{-3} \times 50 \times 10^{-3} \times 2{,}320/(4 \times 920) \\ = 3.96 \times 10^{-3}\,\mathrm{m^3/s} \tag{PI.15}$$

Answer to Problem I.8

The cross-sectional mean velocity, v_m, is defined as

$$v_m = \int_A v dA/A = 2\pi \int_0^R rvdr/(\pi R^2) \qquad (PI.16)$$

where $A(=\pi R^2)$ is the cross-sectional area of the pipe. Substituting Eq. (2.68) into Eq. (pI.16) gives

$$v_m = \int_0^R 2rv_{cl}(1 - r^2/R^2)dr/(R^2) \qquad (PI.17)$$

The following new variable is introduced:

$$x = r/R \qquad (PI.18)$$

Differentiation of Eq. (pI.18) yields

$$dx = dr/R \qquad (PI.19)$$

Equation (pI.17) is transformed into

$$v_m = \int_0^1 2v_{cl}x(1 - x^2)dx \qquad (PI.20)$$

Integration of Eq. (pI.20) gives

$$v_m = 2v_{cl}\int_0^1 (x - x^3)dx = 2v_{cl}\left(\frac{1}{2} - \frac{1}{4}\right) = \frac{v_{cl}}{2} \qquad (PI.21)$$

Problem I.11 Calculate the Reynolds number, given $D = 10$ mm, $v_m = 15$ m/s, and kinematic viscosity of air $\nu_a = 1.5 \times 10^{-5}$ m^2/s.

Problem I.12 Calculate the cross-sectional mean velocity of water flow in a pipe, the Reynolds number, the friction factor, and the pressure drop. The pipe diameter is 50 mm, the pipe length is 20 m, the water flow rate is 1.00×10^{-3} m^3/s, the density of water is 997 kg/m^3, and the kinematic viscosity is 1.00×10^{-6} m^2/s.

Problem I.13 Obtain the dynamic viscosity, μ, given that $D = 1.0$ mm, $L = 1.00$ m, mass flow rate of oil $\rho\ Q_L = 0.260$ kg/h, pressure drop, $\Delta p = 640$ k Pa, and $\rho = 920$ kg/m^3.

Answer to Problem I.11

The Reynolds number, Re, is given by

$$Re = v_m D/\nu_a = 15 \times 10 \times 10^{-3}/(1.5 \times 10^{-5}) = 1.0 \times 10^4 \qquad (PI.22)$$

Answer to Problem I.12

The cross-sectional mean velocity, v_m, is given by

$$v_m = 4Q_L/(\pi D^2) = 4 \times 1.00 \times 10^{-3}/[3.14 \times (50 \times 10^{-3})^2] \\ = 0.510\,\mathrm{m/s} \quad \text{(PI.23)}$$

The Reynolds number, Re, is given by

$$Re = v_m D/\nu = 0.510 \times 50 \times 10^{-3}/(1.00 \times 10^{-6}) = 2.55 \times 10^4 \quad \text{(PI.24)}$$

The friction factor, λ, is calculated from the Blasius equation as

$$\lambda = 0.3164\, Re^{-1/4} = 0.3164 \times (2.55 \times 10^4)^{-1/4} = 0.0250 \quad \text{(PI.25)}$$

The pressure drop, Δp, becomes

$$\Delta p = \lambda(L/D)\rho {v_m}^2/2 = 0.0250 \times [20/(50 \times 10^{-3})] \times 997 \times (0.510)^2/2 \\ = 1.30\,\mathrm{kPa} \quad \text{(PI.26)}$$

where L is the pipe length.

Answer to Problem I.13

The pressure loss is expressed by

$$\Delta p = \lambda \frac{L}{D} \frac{\rho {v_m}^2}{2} \quad \text{(PI.27)}$$

First, the flow in the pipe is assumed to be laminar. The friction factor, λ, is expressed as

$$\lambda = 64/Re = 64\nu_f/(v_m D) \quad \text{(PI.28)}$$

Substituting Eq. (pI.28) into Eq. (pI.27) yields

$$\Delta p = 32\mu L v_m/D^2 \quad \text{(PI.29)}$$

Eq. (pI.29) is rewritten by

$$\mu = \Delta p D^2/(32 L v_m) \quad \text{(PI.30)}$$

The flow rate, Q_L, is given by

$$Q_L = 0.26/(3{,}600 \times 920) = 7.86 \times 10^{-8}\,\mathrm{m^3/s}, \quad \text{(PI.31)}$$

The cross-sectional mean velocity, v_m, is calculated as follows:

$$v_m = 4Q_L/(\pi D^2) = 4 \times 7.86 \times 10^{-8}/[3.14 \times (1.0 \times 10^{-3})^2] \\ = 0.100\,\mathrm{m/s} \quad \text{(PI.32)}$$

Substituting Eq. (pI.32) into Eq. (pI.30) gives

$$\mu = \Delta p D^2/(32 L v_m) = 640 \times 10^3 \times (1.0 \times 10^{-3})^2/(32 \times 1.0 \times 0.100) \\ = 0.200\,\mathrm{Pa\,s} \quad \text{(PI.33)}$$

It is necessary to check whether the laminar flow assumption is adequate or not:

$$Re = \rho v_m D/\mu = 920 \times 0.100 \times 1.0 \times 10^{-3}/0.200 = 0.46 \quad \text{(PI.34)}$$

Accordingly, the laminar flow assumption is adequate.

Problem I.14 A sphere of diameter, D_{pR}, of 1.00 m is immersed in a river. The velocity of the approaching flow, V_R, is 2.0 m/s. A model experiment is carried out using air as the working fluid to predict the drag force acting on the sphere. The diameter of the model sphere, D_{pM}, is 0.10 m. Determine the approaching velocity of air flow, V_M, in terms of the Reynolds number similitude by assuming that the temperatures of air and water are 20 °C. (Ans: 300 m/s)

Problem I.15 A sphere of diameter, D_p, of 5.0 cm is moving in still water at a velocity, V, of 2.0 m/s. The kinematic viscosity of water, ν_w, is 1.01×10^{-6} m^2/s. The drag coefficient, C_D, is 0.44. Determine the hydrodynamic drag acting on the sphere. (Ans: 1.72 N)

Problem I.16 A sphere of density of 3,000 kg/m^3 and diameter of 5.0 cm is falling in still water of density of 998 kg/m^3. The drag coefficient is 0.44. Determine the following quantities:

(a) Volume of sphere, V_p
(b) Buoyancy force acting on sphere, F_B
(c) Weight of sphere, F_w
(d) Drag force acting on sphere, F_D
(e) Terminal velocity of sphere, v_∞ (Ans: 1.72 m/s)

Problem I.17 A man weighing 620 N descends in the atmosphere by parachute. The diameter and weight of the parachute are 5.8 m and 16.8 N, respectively. The density of air is 1.205 kg/m^3. The drag coefficient is assumed to be 1.0. Determine the terminal velocity, v_∞.

Problem I.18 Calculate the cross-sectional mean velocity, v_m, Reynolds number, Re, friction coefficient, λ, and pressure drop, Δp, of water flow given that $D = 40$ mm, $L = 30$ m, $Q_L = 1.00 \times 10^{-3}$ m^3/s, and $\nu = 1.00 \times 10^{-6}$ m^2/s. The critical Reynolds number, Re_c, is 2,320.

Problem I.19 A sphere of diameter of 3.0 cm is moving in still water at a terminal velocity of 3.0 m/s. The density and kinematic viscosity of water are 997 kg/m^3 and 1.01×10^{-6} m^2/s, respectively. Determine the hydrodynamic drag acting on the sphere.

Answer to Problem I.14

According to the Reynolds similitude, the following relationship holds:

$$V_R D_{pR}/\nu_R = V_M D_{pM}/\nu_M \tag{PI.35}$$

$$\begin{aligned} V_M &= V_R\left(D_{pR}/D_{pM}\right)(\nu_M/\nu_R) = 2.0 \times (1.00/0.10) \times [1.50 \times 10^{-5}/(1.0 \times 10^{-6})] \\ &= 300\,\mathrm{m/s} \end{aligned} \tag{PI.36}$$

where the subscripts R and M denote the real process and model, respectively.

Answer to Problem I.15

The density of water, ρ, is 997 kg/m^3. The hydrodynamic drag is expressed by

$$\begin{aligned} F_D &= C_D A_p \rho V^2/2 = 0.44 \times [\pi \times (5.0 \times 10^{-2})^2/4] \times 997 \times (2.0)^2/2 \\ &= 1.72\,\mathrm{N} \end{aligned} \tag{PI.37}$$

Answer to Problem I.16

(a) The volume of the sphere, V_p, is

$$V_p = \pi D_p{}^3/6 = 3.14 \times (5.0 \times 10^{-2})^3/6 = 6.54 \times 10^{-5}\,\mathrm{m}^3 \tag{PI.38}$$

(b) The buoyancy force acting on the sphere, F_B, is

$$F_B = \rho_w V_p g = 998 \times 6.54 \times 10^{-5} \times 9.80 = 0.640\,\mathrm{N} \tag{PI.39}$$

(c) The weight of the sphere, F_w, is

$$Fw = \rho_p V_p g = 3{,}000 \times 6.54 \times 10^{-5} \times 9.80 = 1.923\,\mathrm{N} \tag{PI.40}$$

(d) The drag force acting on the sphere, F_D, is calculated as

$$F_D = F_w - F_B = 1.923 - 0.640 = 1.283\,\mathrm{N} \tag{PI.41}$$

(e) The terminal velocity of the sphere, v_∞, can be calculated from the following relation:

$$v_\infty = [2F_D/(C_D A_p \rho_w)]^{1/2} \tag{PI.42}$$

$$A_p = \pi D_p{}^2/4 = 3.14 \times (5.0 \times 10^{-2})^2/4 = 1.96 \times 10^{-3}\,\mathrm{m}^2 \tag{PI.43}$$

Substituting Eq. (pI.43) and the other quantities into Eq. (pI.42) gives

$$v_\infty = [2 \times 1.283/(0.44 \times 1.96 \times 10^{-3} \times 998]^{1/2} = 1.72\,\mathrm{m/s}) \tag{PI.44}$$

Answer to Problem I.17

The hydrodynamic drag is given by

$$F_D = 620 + 16.8 = 636.8\,\mathrm{N} \tag{PI.45}$$

The terminal velocity can be calculated as follows:

$$A_p = \pi D^2/4 = 3.14 \times (5.8)^2/4 = 26.4\,\mathrm{m}^2 \tag{PI.46}$$

$$\begin{aligned} v_\infty &= [2F_D/(C_D A_p \rho_w)]^{1/2} = [2 \times 636.8/(1.0 \times 26.4 \times 1.205)]^{1/2} \\ &= 6.33\,\mathrm{m/s} \end{aligned} \tag{PI.47}$$

Answer to Problem I.18

$$v_m = Q/A = 1.00 \times 10^{-3}/[\pi \times (0.040)^2/4] = 0.796\,\mathrm{m/s} \tag{PI.48}$$

$$Re = v_m D/\nu = 0.796 \times 0.040/(1.00 \times 10^{-6}) = 3.18 \times 10^4 \tag{PI.49}$$

$$\lambda = 0.3164\mathrm{Re}^{-1/4} = 0.3164\,(3.18 \times 10^4)^{-1/4} = 0.0237 \tag{PI.50}$$

$$\begin{aligned} \Delta p &= \lambda (L/D) \rho v_m{}^2/2 = 0.0237 \times (30/0.040) \times 997 \times (0.796)^2/2 \\ &= 8.77 \times 10^3\,\mathrm{Pa} \end{aligned} \tag{PI.51}$$

Answer to Problem I.19

The Reynolds number is

$$Re = V_\infty D/\nu_w = 3.0 \times 3.0 \times 10^{-2}/(1.01 \times 10^{-6}) = 8.91 \times 10^4 \tag{PI.52}$$

The drag coefficient follows Newton's resistance law, and hence,

$$C_D = 0.44 \tag{PI.53}$$

The hydrodynamic drag is given by

$$\begin{aligned} F_D &= C_D A_p \rho_w V_\infty{}^2/2 \\ &= 0.44 \times [3.14 \times (0.03)^2/4] \times 997 \times (3)^2/2 = 1.39\,\mathrm{N} \end{aligned} \tag{PI.54}$$

Part II

Problem II.1 Derive Eq. (5.17).

Problem II.2 The thickness and thermal conductivity of a flat wall made of concrete are 10 mm and 0.1 W/m K, respectively. The difference between the temperatures of the inner and outer surfaces of the wall is 35 °C. Calculate the heat transfer by conduction in 2 h assuming that the wall surface area is 20 m^2.

Problem II.3 The outer surface of a flat plate made of iron is covered with an insulator. The thickness of the plate is 2.0 cm and the insulator is 5.0 cm thick. The inner wall of the plate is exposed to a gas at 400 °C and the outer surface of the insulator is exposed to the atmosphere at 20 °C. The heat-transfer coefficients on the flat plate side and insulator side are 5.0 and 1.0 W/m^2 K, respectively. The thermal conductivities of the iron plate and the insulator are 40.0 and 0.050 W/m K. Calculate the temperatures on the inner wall of the flat plate and on the outer wall of the insulator and the heat flux.

Answer to Problem II.2

The heat transfer by conduction, Q, is given by

$$\begin{aligned} Q &= \lambda[(\theta_1 - \theta_2)/\delta]At = 0.1 \times [35/0.01] \times 20 \times 3{,}600 \times 2 \\ &= 5.04 \times 10^7\,\mathrm{J} \end{aligned} \tag{PII.1}$$

Answer to Problem II.3

The heat flux, q, is expressed by

$$q = (\theta_a - \theta_b)/(1/\alpha_1 + \delta_1/\lambda_1 + \delta_2/\lambda_2 + 1/\alpha_2) \tag{PII.2}$$

Substituting the abovementioned values into Eq. (pII.2) yields

$$\begin{aligned} q &= (400 - 20)/(1/5 + 0.02/40 + 0.05/0.05 + 1/1) \\ &= 380/(0.2 + 0.0005 + 1 + 1) = 380/2.2005 \\ &= 172.7\,\mathrm{W/m^2} \end{aligned} \tag{PII.3}$$

The wall temperatures are given by

$$\theta_1 = \theta_a - q/\alpha_1 = 400 - 172.7/5 = 365.5\ °\mathrm{C} \tag{PII.4}$$

$$\theta_2 = \theta_b + q/\alpha_2 = 20 + 172.7/1 = 192.7\ °\mathrm{C} \tag{PII.5}$$

Problem II.4 The surfaces of a flat plate 2.0 cm thick are kept at 300 °C and 10 °C. Calculate the heat transfer in 2 h and heat flux assuming that the thermal conductivity is 40 W/m K.

Problem II.5 The difference between the inner and outer surface temperatures of a 50 mm thick pressure vessel is 40 °C. The total surface area, A, of the vessel is 50 m^2. The vessel is constructed of a material having thermal conductivity, λ, of 50 W/m K. Calculate the heat flux and the heat loss in 1 s.

Problem II.6 A flat iron plate 4.0 cm thick is covered with insulator 4.0 cm thick. The inner wall of the plate is exposed to a gas at 350 °C, while the outer wall of the insulator is surrounded by air at a temperature of 20 °C. The heat-transfer coefficients on the inner and outer sides are 5.0 and 2.0 W/m^2 K, respectively. The thermal conductivities of the plate material and insulator are 40.0 and 0.050 W/m K, respectively. Calculate the surface temperatures of the plate and insulator and the heat flux.

Problem II.7 The thickness and thermal conductivity of a flat wall made of concrete are 10 mm and 0.15 W/m K, respectively. The difference in temperature between the inner and outer surfaces of the wall is 40 °C. Calculate the heat transferred by conduction in 3 h assuming that the surface area of the wall is 15 m^2.

Problem II.8 Two plates made of different materials, A and B, have thermal conductivities λ_1 and λ_2, respectively. The plates have equal width, m, but different thicknesses, δ_1 and δ_2. As shown in Fig. PII.1a, b, two types of multilayer plates are constructed using four plates made of A and three plates made of B. Calculate the overall thermal conductivities of the two multilayer plates assuming that the surface temperatures are $\theta_1, \theta_2 (\theta_1 > \theta_2)$.

Answer to Problem II.4

$$\begin{aligned} Q &= \lambda[(\theta_1 - \theta_2)/\delta]At = 40 \times [(300 - 10)/0.02] \times 1 \times 2 \times 3{,}600 \\ &= 4.18 \times 10^9\ \mathrm{J} \end{aligned} \tag{PII.6}$$

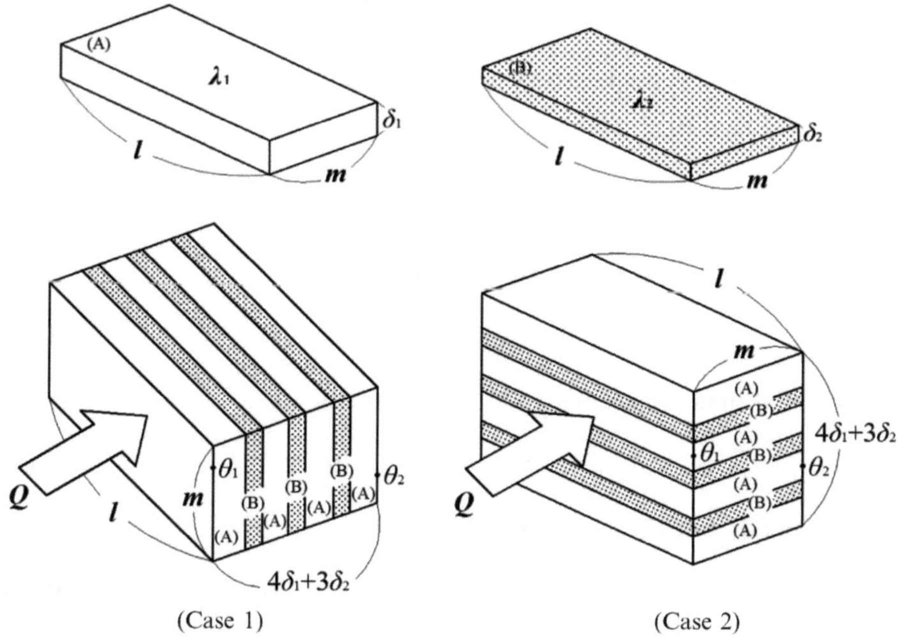

Fig. PII.1 Multilayer plates

Answer to Problem II.5

The wall is regarded as a flat plate. The heat transferred by conduction is given by

$$Q = \lambda[(\theta_1 - \theta_2)/\delta]At = 50 \times (40/0.050) \times 50 \times 1 \\ = 2.00 \times 10^6 \text{ J} \tag{PII.7}$$

Answer to Problem II.6

The heat flux, q, is expressed by

$$\begin{aligned} q &= (\theta_a - \theta_b)/(1/\alpha_1 + \delta_1/\lambda_1 + \delta_2/\lambda_2 + 1/\alpha_2) \\ &= (350 - 20)/(1/5 + 0.04/40 + 0.04/0.05 + 1/2) \\ &= 330/(0.2 + 0.001 + 0.8 + 0.5) = 330/1.501 \\ &= 219.9 \text{ W/m}^2 \end{aligned} \tag{PII.8}$$

The wall temperatures are given by

$$\theta_1 = \theta_a - q/\alpha_1 = 350 - 219.9/5 = 306.0\ ^\circ\text{C} \tag{PII.9}$$

$$\theta_2 = \theta_b + q/\alpha_2 = 20 + 219.9/2 = 130.0\ ^\circ\text{C} \tag{PII.10}$$

Answer to Problem II.7

$$Q = \lambda[(\theta_1 - \theta_2)/\delta]At = 0.15 \times [40/0.010] \times 15 \times 3 \times 3,600 = 9.72 \times 10^7 \text{ J} \quad \text{(PII.11)}$$

Answer to Problem II.8

Case 1 The heat flux, q, can be expressed by

$$q = (\theta_1 - \theta_2)/(4\delta_1/\lambda_1 + 3\delta_2/\lambda_2) \quad \text{(PII.12)}$$

The overall thermal conductivity, λ, is defined as follows:

$$q = \lambda(\theta_1 - \theta_2)/(4\delta_1 + 3\delta_2) \quad \text{(PII.13)}$$

Combining Eqs. (pII.12) and (pII.13) yields

$$\lambda = \lambda_1\lambda_2(4\delta_1 + 3\delta_2)/(4\delta_1\lambda_2 + 3\delta_2\lambda_1) \quad \text{(PII.14)}$$

Case 2 The heat flux for plate A is expressed by

$$q_A = \lambda_1(\theta_1 - \theta_2)/m \quad \text{(PII.15)}$$

In the same manner the heat flux for plate B is given by

$$q_B = \lambda_2(\theta_1 - \theta_2)/m \quad \text{(PII.16)}$$

Accordingly, the total heat transfer, Q, becomes

$$Q = q_A \times 4(\delta_1 l) + q_B \times 3(\delta_2 l) = (4\lambda_1\delta_1 + 3\lambda_2\delta_2)(\theta_1 - \theta_2)l/m \quad \text{(PII.17)}$$

The overall thermal conductivity, λ, is defined as

$$Q = \lambda(\theta_1 - \theta_2)(4\delta_1 + 3\delta_2)l/m \quad \text{(PII.18)}$$

Combination of Eqs. (pII.17) and (pII.18) yields

$$\lambda = (4\lambda_1\delta_1 + 3\lambda_2\delta_2)/(4\delta_1 + 3\delta_2) \quad \text{(PII.19)}$$

Problem II.9 Two materials, 1 and 2, used for reducing heat loss have thermal conductivities λ_1 and λ_2 ($\lambda_2 > \lambda_1$), respectively. Two types of insulation patterns are

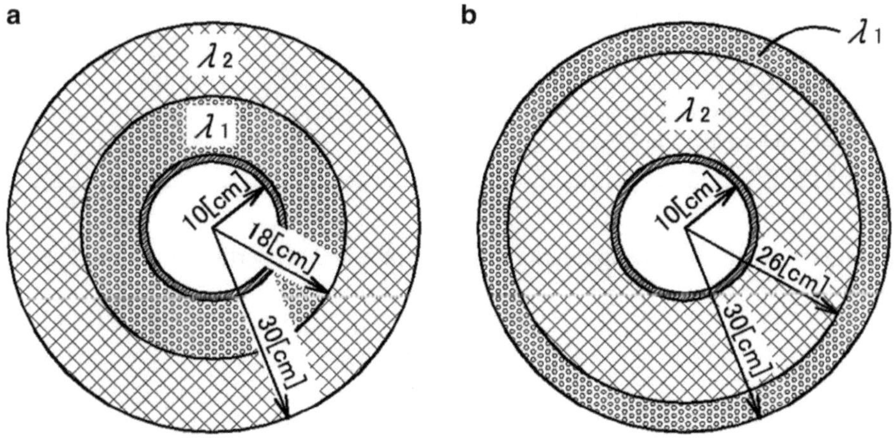

Fig. PII.2 Insulation pattern effect on heat loss

chosen as shown in Fig. PII.2a, b. The amount of the material 1 in (a) is the same as that in (b). The same is true for material 2. Which pattern is more effective for reducing heat loss?

Problem II.10 Steam flows in a hollow cylinder of outer diameter 110 mm and length 30 m. The cylinder is placed in the atmosphere and covered with insulator to reduce heat loss from the surface. The thickness and the thermal conductivity of the insulator are 5 cm and 0.12 W/m K, respectively. The inner and outer surfaces of the insulator are maintained at 100 °C and 30 °C, respectively. Calculate the heat loss from the outer surface of the insulator per unit time (1 s).

Problem II.11 A long hollow cylinder of inner diameter 200 mm and thickness 10 mm is placed vertically in the atmosphere. The temperature of gas flowing in the cylinder is 400 °C and that of air is 20 °C. The heat-transfer coefficients on the inside and outside of the cylinder are 50.0 and 5.0 W/m^2 K, respectively. The thermal conductivity of the cylinder is 40.0 W/m K.

(a) Draw the cylinder with symbols, if necessary.
(b) Calculate the heat transferred per unit length (1 m), i.e., the heat-transfer rate.
(c) Calculate the temperature on the inner and outer walls of the cylinder.

Problem II.12 A hollow sphere has inner radius r_1 and outer radius r_2. The temperatures on the inner and outer walls are kept constant at θ_1 and θ_2 ($\theta_1 > \theta_2$), respectively. Calculate the heat transfer in the radial direction per unit time using Fourier's law.

Problem II.13 A sphere of 6 cm diameter is placed in the atmosphere. The temperature of the sphere is 317 °C and that of air is 17 °C. Calculate the heat loss due to natural convection. The Nusselt number and Grashof number are

Table PII.1 Physical properties of air

Temperature, θ (K)	300	400	600	800
Density, ρ (kg/m^3)	1.176	0.882	0.588	0.441
Kinematic viscosity, ν_f (mm^2/s)	15.83	26.39	52.36	84.5
Thermal conductivity, λ (mW/m K)	26.14	33.05	45.6	56.9
Prandtl number, Pr (–)	0.717	0.715	0.710	0.719

expressed by the following relations and the physical properties of air are given in Table PII.1:

$$Nu = 2 + 0.60 Pr^{1/3} Gr^{1/4} \tag{PII.20}$$

$$Gr = g\beta d^3 (\theta_s - \theta_f)/\nu_f{}^2 \tag{PII.21}$$

Problem II.14 A sphere of radius 6 cm is placed in air flowing at velocity 20 m/s. The temperatures of the sphere and air are 317 °C and 17 °C, respectively. Calculate the heat loss due to forced convection. The Nusselt number is expressed by the following relation and the physical properties of air are given in Table PII.1:

$$Nu = 2 + 0.60 Pr^{1/3} Re^{1/2} \tag{PII.22}$$

Problem II.15 Explain the following:

(a) Three basic types of heat transfer
(b) Reynolds number (definition and physical meaning)
(c) Prandtl number (definition and physical meaning)
(d) Nusselt number (definition and physical meaning)
(e) Fourier's law of conduction
(f) Newton's law of cooling
(g) Natural convection
(h) Forced convection

Answer to Problem II.9

The temperature on the inner wall of the inner layer and that on the outer wall of the outer layer in case (a) are denoted by θ_1 and θ_2, respectively. In the same manner, those in case (b) are denoted by θ_1' and θ_2', respectively. The heat transfer per unit time in the two cases (a) and (b) are expressed by

$$Q_a = 2\pi(\theta_1 - \theta_2)/[(1/\lambda_1)\ln(18/10) + (1/\lambda_2)\ln(30/18)] \tag{PII.23}$$

$$Q_b = 2\pi(\theta_1' - \theta_2')/[(1/\lambda_2)\ln(26/10) + (1/\lambda_1)\ln(30/26)] \tag{PII.24}$$

The difference between the two denominators is expressed by

$$\begin{aligned}&(1/\lambda_1)[\ln(18/10) - \ln(30/26)] + (1/\lambda_2)[\ln(30/18) - \ln(26/10)]\\&= (1/\lambda_1)\ln[18 \times 26/(10 \times 30)] - (1/\lambda_2)\ln[26 \times 18/(30 \times 10)]\\&= (1/\lambda_1)\ln[156/100] - (1/\lambda_2)\ln[156/100]\\&= (1/\lambda_1 - 1/\lambda_2)\ln(156/100)\end{aligned} \quad \text{(PII.25)}$$

This value is positive because $\lambda_2 > \lambda_1$. If we consider the case that $(\theta_1 - \theta_2) = (\theta_1{}' - \theta_2{}')$, $Q_a < Q_b$. Consequently, (a) is more effective for reducing the heat loss.

Answer to Problem II.10

$$\theta_1 = 100\ °\text{C} \quad \text{(PII.26)}$$

$$\theta_2 = 30\ °\text{C} \quad \text{(PII.27)}$$

$$r_1 = 110/2 = 55\ \text{mm} \quad \text{(PII.28)}$$

$$r_2 = 55 + 50 = 105\ \text{mm} \quad \text{(PII.29)}$$

$$\begin{aligned}\theta &= 2\pi l\lambda(\theta_1 - \theta_2)/\ln(r_2/r_1)\\&= 2 \times 3.14 \times 30 \times 0.12 \times (100 - 30)/\ln(0.105/0.055)\\&= 1.58 \times 10^3/0.647 = 2.44 \times 10^3\ \text{W}\end{aligned} \quad \text{(PII.30)}$$

Answer to Problem II.11

$$\theta_a = 400\ °\text{C} \quad \text{(PII.31)}$$

$$\theta_b = 20\ °\text{C} \quad \text{(PII.32)}$$

$$\alpha_1 = 50.0\ \text{W/m}^2\ \text{K} \quad \text{(PII.33)}$$

$$\alpha_2 = 5.0\ \text{W/m}^2\ \text{K} \quad \text{(PII.34)}$$

$$\lambda = 40.0\ \text{W/m K} \quad \text{(PII.35)}$$

$$r_1 = 200/2 = 100\ \text{mm} \quad \text{(PII.36)}$$

$$r_2 = r_1 + 10 = 100 + 10 = 110\ \text{mm} \quad \text{(PII.37)}$$

The heat-transfer rate is given by

$$\begin{aligned} Q &= 2\pi l(\theta_a - \theta_b)/[1/(\alpha_1 r_1) + (1/\lambda)\ln(r_2/r_1) + 1/(\alpha_2 r_2)] \\ &= 2 \times 3.14 \times 1 \times (400 - 20)/[1/(50.0 \times 0.100) + (1/40.0)\ln(0.110/0.100) \\ &\quad + 1/(5.0 \times 0.110)] \\ &= 2386.3/2.02 = 1.18 \times 10^3 \text{ W} \end{aligned} \tag{PII.38}$$

The temperatures on the inner and outer walls are expressed by

$$\begin{aligned} \theta_1 &= \theta_a - Q/(2\pi r_1 \alpha_1) = 400 - 1181/(2 \times 3.14 \times 0.100 \times 50.0) \\ &= 400 - 37.62 = 362.38\ ^\circ\text{C} \end{aligned} \tag{PII.39}$$

$$\begin{aligned} \theta_2 &= \theta_b + Q/(2\pi r_2 \alpha_2) = 20 + 1{,}181/(2 \times 3.14 \times 0.110 \times 5.0) \\ &= 20 + 341.92 = 361.92\ ^\circ\text{C} \end{aligned} \tag{PII.40}$$

Answer to Problem II.12

Fourier's law is expressed by

$$Q = -\lambda(d\theta/dr)A \tag{PII.41}$$

$$A = 4\pi r^2 \tag{PII.42}$$

Substituting Eq. (pII.42) into Eq. (pII.41) yields

$$Q = -4\pi\lambda r^2(d\theta/dr) \tag{PII.43}$$

$$d\theta = -Q/(4\pi\lambda)dr/r^2 \tag{PII.44}$$

The boundary conditions are

$$\theta = \theta_1 \text{ at } r = r_1 \tag{PII.45}$$

$$\theta = \theta_2 \text{ at } r = r_2 \tag{PII.46}$$

Integrating Eq. (pII.44) subject to the boundary conditions of Eqs. (pII.45) and (pII.46) yields

$$Q = 4\pi\lambda(\theta_1 - \theta_2)/[1/r_1 - 1/r_2] \tag{PII.47}$$

Answer to Problem II.13

The mean (film) temperature is given by

$$\theta_{fm} = (317 + 17)/2 = 167\,°C \tag{PII.48}$$

The physical properties of air at this temperature are calculated from Table PII.1 as follows:
Kinematic viscosity, ν_f

$$\begin{aligned}\nu_f &= 26.39 + 40 \times (52.36 - 26.39)/(600 - 400) = 26.39 + 5.19 \\ &= 31.58\,\mathrm{mm^2/s} = 31.58 \times 10^{-6}\,\mathrm{m^2/s}\end{aligned} \tag{PII.49}$$

Thermal conductivity of air, λ_f

$$\begin{aligned}\lambda_f &= 33.05 + 40 \times (45.60 - 33.05)/(600 - 400) = 33.05 + 2.51 \\ &= 35.56\,\mathrm{m\,W/m\,K} = 35.56 \times 10^{-3}\,\mathrm{W/m\,K}\end{aligned} \tag{PII.50}$$

Prandtl number, Pr

$$\begin{aligned}Pr &= 0.715 + 40 \times (0.710 - 0.715)/(600 - 400) = 0.715 - 0.001 \\ &= 0.714\end{aligned} \tag{PII.51}$$

The volume coefficient of expansion, β, is given by

$$\beta = 1/(273 + 17) = 1/290 \tag{PII.52}$$

The Grashof and Nusselt numbers are calculated as follows:

$$\begin{aligned}Gr &= g\beta d^3(\theta_s - \theta_f)/\nu_f^2 \\ &= 9.80 \times (1/290) \times (0.06)^3 \times (317 - 17)/(31.58 \times 10^{-6})^2 \\ &= 2.196 \times 10^6\end{aligned} \tag{PII.53}$$

$$\begin{aligned}Nu &= 2 + 0.60 Pr^{1/3} Gr^{1/4} \\ &= 2 + 0.60 \times (0.714)^{1/3} \times (2.196 \times 10^6)^{1/4} \\ &= 2 + 0.60 \times 0.8938 \times 38.49 = 2 + 20.64 = 22.64[-]\end{aligned} \tag{PII.54}$$

The heat-transfer coefficient, α, is calculated as

$$\begin{aligned}\alpha &= Nu\lambda_f/d \\ &= 22.64 \times 35.56 \times 10^{-3}/0.06 = 13.42\,\mathrm{W/m^2\,K}\end{aligned} \tag{PII.55}$$

Accordingly, the heat transferred by natural convection per unit time is given by

$$\begin{aligned} Q &= \alpha(\theta_s - \theta_f)A = \alpha(\theta_s - \theta_f)\pi d^2 \\ &= 13.42 \times (317 - 17) \times \pi \times (0.06)^2 = 45.53\,\text{W} \end{aligned} \quad \text{(PII.56)}$$

Answer to Problem II.14

The diameter of the sphere, d, is

$$d = 0.06 \times 2 = 0.12\,\text{m} \quad \text{(PII.57)}$$

The kinematic viscosity, ν_f, thermal conductivity, λ_f, and Prandtl number, Pr, are

$$\nu_f = 31.58 \times 10^{-6}\,\text{m}^2/\text{s} \quad \text{(PII.58)}$$

$$\lambda_f = 35.56 \times 10^{-3}\,\text{W/m K} \quad \text{(PII.59)}$$

$$Pr = 0.714 \quad \text{(PII.60)}$$

The Reynolds number, Re, Nusselt number, Nu, and heat-transfer coefficient, α, are given by

$$Re = Vd/\nu_f = 20 \times 0.12/(31.58 \times 10^{-6}) = 7.60 \times 10^4\,[-] \quad \text{(PII.61)}$$

$$\begin{aligned} Nu &= 2 + 0.60 Pr^{1/3} Re^{1/2} \\ &= 2 + 0.60 \times (0.714)^{1/3} \times (7.60 \times 10^4)^{1/2} \\ &= 2 + 0.60 \times 0.8938 \times 275.7 = 2 + 147.8 = 149.8\,[-] \end{aligned} \quad \text{(PII.62)}$$

$$\begin{aligned} \alpha &= Nu\,\lambda_f/d \\ &= 149.8 \times 35.56 \times 10^{-3}/0.12 = 44.40\,\text{W/m}^2\,\text{K} \end{aligned} \quad \text{(PII.63)}$$

The heat-transfer rate by natural convection is given by

$$\begin{aligned} Q &= \alpha(\theta_s - \theta_f)A = \alpha(\theta_s - \theta_f)\pi d^2 \\ &= 44.40 \times (317 - 17) \times \pi \times (0.12)^2 = 602.6\,\text{W} \end{aligned} \quad \text{(PII.64)}$$

Problem II.16 The temperature in a room is 30 °C and that in the atmosphere is 5 °C. The thickness of a window made of glass is 4.0 mm. The thermal conductivity of glass, λ, is 0.76 W/m K, and the heat-transfer coefficients inside and outside the room, α_1 and α_2, are 10 and 50 W/m^2 K, respectively. Calculate the heat flux, q, defined as heat flow per unit time and unit area. (Ans: 199.5 J/m^2 s)

Problem II.17 Two parallel plates are placed vertically at a distance L apart. The temperatures of the plates are θ_1 and θ_2 ($\theta_1 > \theta_2$). Calculate heat-transfer rates, Q_i and Q_{ii}, in the following two cases:

(i) Two cylindrical rods of cross-sectional area, A, and length, L, connect the two plates separately. The thermal conductivity of one rod is λ_a and that of the other rod is λ_b. Consider one-dimensional heat conduction, that is, the heat losses on the surfaces of the rods are negligible.
(ii) Two cylindrical rods of cross-sectional area, 2A, and length, L/2, are combined into one cylindrical rod and then used to connect the two plates. The thermal conductivity of one rod is λ_a and that of the other rod is λ_b. Consider one-dimensional heat conduction.

Which heat-transfer rate is higher, Q_i or Q_{ii}?

Answer to Problem II.16

The heat flux, q, can be calculated from Eq. (5.16) to give

$$\begin{aligned} q &= (\theta_a - \theta_b)/(1/\alpha_1 + \delta/\lambda + 1/\alpha_2) \\ &= (30 - 5)/(1/10 + 0.004/0.76 + 1/50) \\ &= 25/(0.1 + 0.005263 + 0.02) = 25/0.1253 = 199.5\,\mathrm{W/m^2} \end{aligned} \quad \text{(PII.65)}$$

Answer to Problem II.17

Two heat-transfer rates can be evaluated in the following manner.

Case 1 The heat-transfer rates through the cylindrical rods 1 and 2 are expressed by

$$Q_a = \lambda_a[(\theta_1 - \theta_2)/L)]At \quad \text{(PII.66)}$$

$$Q_b = \lambda_b[(\theta_1 - \theta_2)/L)]At \quad \text{(PII.67)}$$

Accordingly, Q_i is given by

$$\begin{aligned} Q_i &= Q_a + Q_b \\ &= (\lambda_a + \lambda_b)[(\theta_1 - \theta_2)/L)]At \end{aligned} \quad \text{(PII.68)}$$

Case 2 The heat-transfer rate, Q_{ii}, is readily derived from Eq. (5.20) as follows:

$$\begin{aligned} Q_{ii} &= (\theta_1 - \theta_2)\ (2At)\ [(L/2)/\lambda_a + (L/2)/\lambda_b] \\ &= [4\lambda_a\lambda_b/(\lambda_a + \lambda_b)]\ [(\theta_1 - \theta_2)/L)]At \end{aligned} \quad \text{(PII.69)}$$

We need to judge which heat-transfer rate is higher, Q_i or Q_{ii}. At a glance, it is sufficient to examine the following function:

$$y = (\lambda_a + \lambda_b) - 4\lambda_a\lambda_b/(\lambda_a + \lambda_b) \quad \text{(PII.70)}$$

Equation (pII.70) can be transformed into

$$\begin{aligned} y &= [(\lambda_a + \lambda_b)^2 - 4\lambda_a\lambda_b]/(\lambda_a + \lambda_b) \\ &= ({\lambda_a}^2 + 2\lambda_a\lambda_b + {\lambda_b}^2 - 4\lambda_a\lambda_b)/(\lambda_a + \lambda_b) \\ &= ({\lambda_a}^2 - 2\lambda_a\lambda_b + {\lambda_b}^2)/(\lambda_a + \lambda_b) \\ &= (\lambda_a - \lambda_b)^2/(\lambda_a + \lambda_b) \end{aligned} \quad \text{(PII.71)}$$

Equation (pII.71) indicates that y is always positive for $\lambda_a \neq \lambda_b$, and we can conclude that Q_i is higher than Q_{ii} under this condition.

Problem II.18 A long hollow cylinder of inner diameter, d_1, of 180 mm and thickness, δ, of 8.0 mm is placed vertically. The temperature of the fluid inside the cylinder, θ_a, is 300 °C and the temperature of the fluid outside, θ_b, is 20 °C. The heat-transfer coefficients inside and outside the cylinder, α_1 and α_2, are 50.0 W/m^2 K and 5.0 W/m^2 K, respectively. The heat conductivity of the cylinder, λ, is 50.0 W/m^2 K. Calculate the heat-transfer rate over 2 m length of cylinder. Determine the inner and outer wall temperatures, θ_1 and θ_2. (Ans: $Q = 1{,}554$ W, $\theta_1 = 272.5$ °C, $\theta_2 = 272.3$ °C)

Problem II.19 A sphere of radius 4.0 cm is placed in the atmosphere. The temperature of the sphere is 427 °C and the surrounding air is at 27 °C. Calculate the heat loss due to natural convection. The Nusselt number and Grashof number are expressed by the following relations, and the physical properties of air are given in Table PII.1. (Ans: 104.3 W)

$$Nu = 2 + 0.60Pr^{1/3}Gr^{1/4} \quad \text{(PII.72)}$$

$$Gr = g\beta d^3(\theta_s - \theta_f)/{\nu_f}^2 \quad \text{(PII.73)}$$

Answer to Problem II.18

The heat-transfer rate per unit length is given by

$$Q/l = 2\pi\lambda(\theta_1 - \theta_2)/[\lambda/(\alpha_1 r_1) + \ln(r_2/r_1) + \lambda/(\alpha_2 r_2)] \quad \text{(PII.74)}$$

where

$$r_1 = 180/2 = 90\,\text{mm} \quad \text{(PII.75)}$$

$$r_2 = 90 + 8.0 = 98\,\text{mm} \quad \text{(PII.76)}$$

Substituting these values and related quantities into Eq. (pII.74) yields

$$Q/l = 2\pi \times 50 \times (300 - 20)/[50/(50.0 \times 0.090) + \ln(0.098/0.090) + 50/(5.0 \times 0.098)] = 87920/113.24 = 776.4\,\mathrm{W/m} \quad \text{(PII.77)}$$

Accordingly,

$$Q = 776.4 \times 2 = 1{,}553\,\mathrm{W} \quad \text{(PII.78)}$$

$$\theta_1 = \theta_a - Q/(\alpha_1 A_1) = 300 - 1{,}553/(50.0 \times 2 \times 3.14 \times 0.090 \times 2) = 300 - 27.47 = 272.5\,^{\circ}\mathrm{C} \quad \text{(PII.79)}$$

$$\theta_2 = \theta_b + Q/(\alpha_2 A_2) = 20 - 1{,}553/(5.0 \times 2 \times 3.14 \times 0.098 \times 2) = 20 + 252.31 = 272.3\,^{\circ}\mathrm{C} \quad \text{(PII.80)}$$

Answer to Problem II.19

The mean temperature, θ_{fm}, is given by

$$\theta_{fm} = (427 + 27)/2 = 227\,^{\circ}\mathrm{C} = 500\,\mathrm{K} \quad \text{(PII.81)}$$

The physical properties can be obtained from Table PII.1 as follows:

$$\nu_f = (26.39 + 52.36)/2 = 39.38\,\mathrm{mm^2/s} = 39.38 \times 10^{-6}\,\mathrm{m^2/s} \quad \text{(PII.82)}$$

$$Pr = (0.715 + 0.710)/2 = 0.713\,[-] \quad \text{(PII.83)}$$

$$\lambda_f = (33.05 + 45.6)/2 = 39.33\,\mathrm{mW/m\,K} = 39.33 \times 10^{-3}\,\mathrm{W/m\,K} \quad \text{(PII.84)}$$

Substituting the above and following values into the definitions of the Grashof number yields Eq. (pII.90):

$$\theta_f = 27 + 273 = 300\,\mathrm{K} \quad \text{(PII.85)}$$

$$\beta = 1/300\,(1/\mathrm{K}) \quad \text{(PII.86)}$$

$$d = 0.08\,\mathrm{m} \quad \text{(PII.87)}$$

$$\theta_s - \theta_f = 427 - 27 = 400\,\mathrm{K} \quad \text{(PII.88)}$$

$$g = 9.8\,\mathrm{m/s^2} \quad \text{(PII.89)}$$

$$Gr = 9.8 \times (1/300) \times (0.08)^3 \times 400/(39.38 \times 10^{-6})^2 = 4.314 \times 10^6 \quad \text{(PII.90)}$$

From Eqs. (pII.83) and (pII.90) we have

$$Pr^{1/3} \cdot Gr^{1/4} = (0.713)^{1/3} \times (4.314 \times 10^6)^{1/4} = 40.65 \quad \text{(PII.91)}$$

As this value is included in the applicable range of Eq. (pII.72), the Nusselt number is calculated from Eq. (pII.72):

$$\begin{aligned} Nu &= 2 + 0.60Pr^{1/3} \cdot Gr^{1/4} \\ &= 2 + 0.60(0.715)^{1/3} \times (4.314 \times 10^6)^{1/4} \\ &= 2 + 0.60 \times 40.65 = 26.39 \end{aligned} \quad \text{(PII.92)}$$

The heat-transfer coefficient, α, is determined from the definition of the Nusselt number:

$$\begin{aligned} \alpha &= 26.39\lambda_f/d = 26.39 \times (39.33 \times 10^{-3})/0.08 \\ &= 12.97\,\text{W/m}^2\,\text{K} \end{aligned} \quad \text{(PII.93)}$$

$$\begin{aligned} Q &= \alpha(\theta_s - \theta_f)A = 12.97 \times (427 - 27) \times 3.14 \times (0.08)^2 \\ &= 104.3\,\text{W} \end{aligned} \quad \text{(PII.94)}$$

Problem II.20 The temperature in a room is 30 °C and the atmosphere is at 5 °C. The thickness of the window made of glass is 4.0 mm. The thermal conductivity of glass, λ, is 0.76 W/m K, and the heat-transfer coefficients inside and outside the room, α_1 and α_2, are 10 and 50 W/m^2 K, respectively. Calculate the heat flux, q, defined as heat flow per unit time and unit area. (Ans: 199.5 J/m^2 s)

Answer to Problem II.20

The heat flux, q, is expressed by

$$\begin{aligned} q &= (\theta_a - \theta_b)/(1/\alpha_1 + \delta/\lambda + 1/\alpha_2) \\ &= (30 - 5)/(1/10 + 0.004/0.76 + 1/50) \\ &= 25/(0.1 + 0.005263 + 0.02) = 25/0.1253 \\ &= 199.5\,\text{W/m}^2 \end{aligned} \quad \text{(PII.95)}$$

Problem II.21 A sphere of diameter 7.0 cm is placed in a flow of air. The temperatures of the sphere and air are 427 °C and 27 °C, respectively. The approaching velocity of air flow is 17 m/s. Calculate the heat loss due to forced convection. The physical properties of air are given in Table PII.1. (Ans: 328.5 W; note that derivation is not given.)

Problem II.22 Derive the following expression for the buoyancy force in Eq. (6.9a):

$$F_{\mathrm{B}} = \rho g d^3 \beta(\theta_{\mathrm{s}} - \theta_{\mathrm{f}}) \tag{PII.96}$$

Answer to Problem II.22

The buoyancy force, F_{B}, is expressed by

$$F_{\mathrm{B}} = V_{\mathrm{d}} \rho g \tag{PII.97}$$

where V_{d} is the volume of fluid displaced, ρ is the density of fluid, and g is the acceleration due to gravity.

The volume of fluid affected by a sphere of diameter, d, can be approximated by d^3 as the volume of the sphere is $\pi d^3/6$. This volume of fluid is expanded due to the heat transferred from the sphere as follows:

$$d^3 \to [d^3 + d^3 \beta(\theta_{\mathrm{s}} - \theta_{\mathrm{f}})] \tag{PII.98}$$

Accordingly, the volume of the fluid displaced, V_{d}, is given by

$$V_{\mathrm{d}} = [d^3 + d^3 \beta(\theta_{\mathrm{s}} - \theta_{\mathrm{f}})] - d^3 = d^3 \beta(\theta_{\mathrm{s}} - \theta_{\mathrm{f}}) \tag{PII.99}$$

Substituting Eq. (pII.99) into Eq. (pII.97) gives

$$F_{\mathrm{B}} = \rho g d^3 \beta(\theta_{\mathrm{s}} - \theta_{\mathrm{f}}) \tag{PII.100}$$

This expression is the same as Eq. (pII.96).

Problem II.23 Derive the following expression for the shearing force in Eq. (6.9):

$$F_\tau = \tau A_{\mathrm{s}} = \mu(v/d)d^2 \tag{PII.101}$$

Answer to Problem II.23

Newton's law of viscosity is described by

$$\tau = \mu dv/dy \tag{PII.102}$$

where τ is the shear stress acting on the surface of the sphere, μ is the dynamic viscosity of fluid, v is the velocity of fluid motion induced by convection around the sphere, and y is the distance measured from the surface of the sphere.

The viscous force, F_τ, is expressed by

$$F_\tau = \tau A_{\mathrm{s}} \tag{PII.103}$$

where A_s is the surface area of the sphere given by

$$A_s = \pi d^2 \rightarrow A_s = d^2 \tag{PII.104}$$

In addition, the velocity gradient can be assumed to be

$$dv/dy = v/d \tag{PII.105}$$

Combining Eqs. (pII.101) and (pII.105) gives

$$\tau = \mu dv/dy = \mu v/d \tag{PII.106}$$

Substituting Eqs. (pII.104) and (pII.106) into Eq. (pII.103) yields

$$F_\tau = \tau A_s = \mu(v/d)d^2 \tag{PII.107}$$

The dynamic viscosity, μ, is expressed by

$$\mu = \rho \nu_f \tag{PII.108}$$

where ν_f is the kinematic viscosity. The viscous force therefore is given by

$$F_\tau = \rho \nu_f (v/d) d^2 \tag{PII.109}$$

Part III

Problem III.1 Helium gas at 25 °C and 4 bars is stored in a spherical Pyrex container of 100 mm inside radius and 10 mm thickness. (a) Calculate the rate of mass loss from the container. (b) If mass flux rate is 8×10^{-15} kg/s, find the pressure at which helium is stored. ($D_{AB} = 0.4 \times 10^{-13}$ m^2/s, $S = 0.45 \times 10^{-3}$ k mol/m^3.bar). (Ans: $P = 8$ bars)

Answer to Problem III.1

Part A

$$N'_{Ar} = \frac{C_{A,S1} - C_{A,S2}}{R_{m,diff}} \tag{PIII.1}$$

$$R_{m,diff} = \frac{1}{4\pi D_{AB}}\left(\frac{1}{R_1} - \frac{1}{R_2}\right) = 1.81 \times 10^{12}\ \frac{s}{m^2} \tag{PIII.2}$$

$$C_{A,S1} = S \times P_A = 0.45 \times 10^{-2} \times 4 = 1.8 \times 10^{-2}\,\text{k mol/m}^3 \quad \text{(PIII.3)}$$

$$C_{A,S2} = 0 \quad \text{(PIII.4)}$$

$$n_{A,r} = m_A \times N_{A,r} \quad \text{(PIII.5)}$$

$$n_{A,r} = 4\frac{\text{kg}}{\text{mol}} \times \frac{1.8 \times 10^{-2}\,\text{k mol/m}^3}{1.81 \times 10^{12}\,\text{s/m}^3} \quad \text{(PIII.6)}$$

$$n_{A,r} = 4 \times 10^{-25}\frac{\text{kg}}{\text{s}} \quad \text{(PIII.7)}$$

Part B

$$n_{A,r} = 8 \times 10^{-25}\frac{\text{kg}}{\text{s}} \quad \text{(PIII.8)}$$

$$N_{A,r} = 2 \times 10^{-15}\frac{\text{k mol}}{\text{s}} \quad \text{(PIII.9)}$$

$$C_{A,S1} = S \times P_A = 0.45 \times 10^{-3}\text{k mol/m}^3 \quad \text{(PIII.10)}$$

$$P_A = \frac{S}{0.45 \times 10^{-3}} \quad \text{(PIII.11)}$$

$$P_A = 8\ \text{bars} \quad \text{(PIII.12)}$$

Problem III.2 Hydrogen flows in a tube at the rate of 8×10^{-6} k mol/s per meter length. The tube is 1 mm thick and has inner diameter of 50 mm. The outer surface is exposed to a gas stream for which the hydrogen partial pressure is 0.1 atm. The mass diffusivity and solubility of hydrogen in the tube material are 1.8×10^{-11} m^2/s and 160 k mol/m^3 $\times$ atm, respectively. Assuming the system is at 500 K, estimate the pressure at which the hydrogen flows within the tube. (Ans: $P = 8.65$ atm)

Answer to Problem III.2

$$m_A = 2\,\text{kg/k mol for } H_2 \quad \text{(PIII.13)}$$

Then,

$$N_{A,r} = 8 \times 10^{-6}\frac{\text{k mol}}{\text{s}} \times \text{m} \quad \text{(PIII.14)}$$

$$N_{A,r} = \frac{2\pi D_{AB}(C_A(r_1) - C_A(r_2))}{\ln\left(\frac{D_O}{D_I}\right)} \quad \text{(PIII.15)}$$

$$8 \times 10^{-6} \frac{\text{k mol}}{\text{s}} \times \text{m} = \frac{2\pi \times 1.8 \times 10^{-11}(C_A(r_1) - C_A(r_2))}{\ln\left(\frac{51}{50}\right)} \quad \text{(PIII.16)}$$

$$C_A(r_1) - C_A(r_2) = 1.4 \times 10^3\,\text{k mol/m}^3 \quad \text{(PIII.17)}$$

$$C_A(r_2) = 160 \times 0.1 = 16\,\text{k mol/m}^3 \quad \text{(PIII.18)}$$

$$C_A(r_1) = 1{,}384\,\text{k mol/m}^3 \quad \text{(PIII.19)}$$

$$P_{A,1} = \frac{C_A(r_1)}{S_{AB}} \quad \text{(PIII.20)}$$

$$P_{A,1} = 8.65\,\text{atm} \quad \text{(PIII.21)}$$

Problem III.3 A rubber plug that is 40 mm thick and has surface area of 500 mm^2 is used to contain CO_2 at 25 °C in a 20 L vessel. If the pressure is dropping at the rate of -9×10^{-10} bar/s, what was the original pressure inside the vessel? Assume

$$D_{AB} = 0.11 \times 10^{-9}\text{m}^2/\text{s} = \text{and } S = 40.15 \times 10^{-3} \frac{\text{k mol}}{\text{m}^3} \times \text{ba}r \quad \text{(PIII.22)}$$

(Ans: $P_{A,1} = 13.15$bars)

Answer to Problem III.3

$$\frac{dP_A}{dt} = -\frac{N_A RT}{V} \quad \text{(PIII.23)}$$

where $V = 2 \times 10^{-2}\ \text{m}^3$ and R is universal gas constant

$$-9 \times 10^{-10} = -\frac{N_A \times 8.314 \times 10^{-2} \times 298}{2 \times 10^{-2}} \quad \text{(PIII.24)}$$

$$N_A = 7.26 \times 10^{-13} \frac{\text{k mol}}{\text{s}} \quad \text{(PIII.25)}$$

$$N_A = AD_{AB} \frac{C_{A,1} - C_{A,2}}{L} \quad \text{(PIII.26)}$$

$$C_{A,1} - C_{A,2} = 0.528 \frac{\text{k mol}}{\text{m}^3} \quad \text{(PIII.27)}$$

$$C_{A,2} = 0 \quad \text{(PIII.28)}$$

$$C_{A,1} = 0.528 \frac{\text{k mol}}{\text{m}^3} \quad \text{(PIII.29)}$$

$$P_{A,1} = \frac{C_{A,1}}{S} \quad \text{(PIII.30)}$$

$$P_{A,1} = 13.15 \text{ bars} \quad \text{(PIII.31)}$$

Part IV

Problem IV.1 Air is injected into a cylindrical water bath of diameter, D, of 400 mm and depth, H_L, of 600 mm through a centered single-hole bottom nozzle of inner diameter, d_{ni}, of 2.0 mm. The nozzle is wetted by water and the air flow rate, Q_g, is 500 cm^3/s. The densities of air and water, ρ_g and ρ_L, are 1.2 kg/m^3 and 997 kg/m^3, respectively. Determine the following quantities:

(a) Frequency of bubble formation, f_{BF}
(b) Modified Froude number defined in Eq. (9.64), Fr_m
(c) Centerline value of gas holdup, α_{cl}, at $z = 300$ mm
(d) Half-value radius of gas holdup, b_α, at $z = 300$ mm
(e) Mean bubble rising velocity, $\bar{u}_B$, at $z = 300$ mm
(f) Centerline value of water velocity, $\bar{u}_{cl}$, at $z = 300$ mm
(g) Half-value radius of water flow velocity, b_u, at $z = 300$ mm

Answer to Problem IV.1

$$\begin{aligned} f_{BF} &= 12.1 Q_g{}^{0.133} d_{ni}{}^{-0.434} = 12.1 \times (500)^{0.133} \times (0.2)^{-0.434} \\ &= 12.1 \times 2.29 \times 2.01 = 55.7\,\text{Hz} \end{aligned} \quad \text{(PIV.1)}$$

$$\begin{aligned} Fr_m &= \rho_g Q_g{}^2/(\rho_L g d_{ni}{}^5) = 1.2 \times (500 \times 10^{-6})^2/[997 \times 9.80 \times (2.0 \times 10^{-3})^5] \\ &= 1.2 \times 2.5 \times 10^{-7}/(3.13 \times 10^{-10}) = 958 \end{aligned} \quad \text{(PIV.2)}$$

Problem IV.2 Estimate the frequency of bubble formation, f_{BF} (Hz), and bubble diameter, d_B, for a poorly wetted single-hole nozzle of inner diameter, d_{ni}, of 2.0 mm and outer diameter of 5.0 mm on the basis of Eq. (9.50). The gas flow rate, Q_g, is 30 cm^3/s.

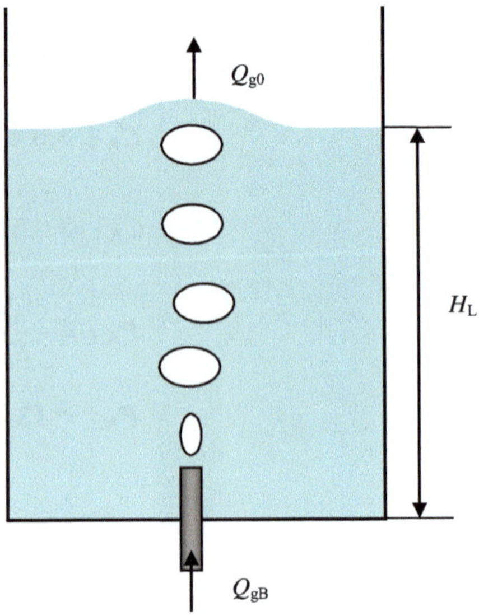

Fig. PIV.1 Deepwater pool subjected to bottom gas injection

Answer to Problem IV.2

The frequency of bubble formation can be calculated from Eq. (9.50) as follows:

$$\begin{aligned} f_{BF} &= 12.1 Q_g{}^{0.133} d_{no}{}^{-0.434} = 12.1 \times (30)^{0.133} \times (0.50)^{-0.434} \\ &= 12.1 \times 1.572 \times 1.35 = 25.8\,\text{Hz} \end{aligned} \tag{PIV.3}$$

The bubble diameter therefore is given by

$$\begin{aligned} d_B &= [6Q_g/(\pi f_{BF})]^{1/3} = 0.54[Q_g d_{no}{}^{0.5}]^{0.289} \\ &= 0.54[30 \times (0.50)^{0.5}]^{0.289} = 1.31\,\text{cm} \end{aligned} \tag{PIV.4}$$

Problem IV.3 Air is injected into a water bath through a single-hole bottom nozzle at a flow rate, Q_g, of 3.0×10^{-4} m^3/s (see Fig. PIV.1). The bath diameter, D, and the inner diameter of the nozzle, d_{ni}, are 2.00×10^{-1} m and 3.0×10^{-3} m, respectively. Calculate the cross-sectional mean air velocity, v_n, and the Mach number, M, in the nozzle. The speed of sound, c, is assumed to be 340 m/s. In addition, calculate the superficial velocity of gas in the bath, j_g, where j_g is defined as follows:

$$j_g = Q_g/(\pi D^2/4) \tag{PIV.5}$$

$$(\text{Ans. } v_n = 42.5\,\text{m/s},\ M = 0.125, j_g = 9.55 \times 10^{-3}\,\text{m/s})$$

Problem IV.4 Air is injected into a water pool through a single-hole bottom nozzle at a flow rate, Q_{gB}, of 3.0×10^{-4} m^3/s (see Fig. 9.10). The depth of the pool, H, is 15 m. Calculate the air flow rate at the surface of the pool, Q_{gS}. The density of water, ρ_w, is 998 kg/m^3 and the atmospheric pressure, p_0, is 101.3 k Pa. The temperature of water, θ, is assumed to be constant everywhere in the bath. The pressure and temperature of the air at the bottom of the pool are assumed to be equal to their respective values for the liquid there. (Ans: $Q_{gS} = 7.34 \times 10^{-4}$ m^3/s)

Answer to Problem IV.3

The mean air velocity, v_n, is calculated in the following manner:

$$\begin{aligned} v_n &= 4Q_g/(\pi d_{ni}{}^2) = 4 \times 3.0 \times 10^{-4}/[3.14 \times (3.0 \times 10^{-3})^2] \\ &= 42.5\,\text{m/s} \end{aligned} \quad \text{(PIV.6)}$$

The Mach number, M, is given by

$$M = v_n/c = 42.5/340 = 0.125 \quad \text{(PIV.7)}$$

The superficial velocity of gas, j_g, is expressed by

$$\begin{aligned} j_g &= 4Q_g/(\pi D^2) = 4 \times 3.0 \times 10^{-4}/[3.14 \times (2.00 \times 10^{-1})^2] \\ &= 9.55 \times 10^{-3}\,\text{m/s} \end{aligned} \quad \text{(PIV.8)}$$

Answer to Problem IV.4

The absolute pressure, p_B, at the bottom of the pool is given by

$$\begin{aligned} p_B &= \rho g H_L + p_0 = 998 \times 9.8 \times 15 + 101.3 \times 10^3 \\ &= 146.7 \times 10^3 + 101.3 \times 10^3 = 248\,\text{k Pa} \end{aligned} \quad \text{(PIV.9)}$$

We assume that the perfect-gas law is valid under this condition:

$$pV = mR\theta \quad \text{(PIV.10)}$$

where p is the absolute pressure of the gas, V is the volume of the gas, m is the mass of the gas, and R is the gas constant. As m, R, and θ are constant under the present condition, Eq. (pIV.10) reduces to

$$pV = \text{const.} \quad \text{(PIV.11)}$$

The volume of gas, V, in this equation can be replaced by gas flow rate, Q_g, as follows:

$$pQ_g = \text{const.} \quad \text{(PIV.12)}$$

The following relationship is obtained because the pressure on the pool surface is p_0:

$$p_B Q_{gB} = p_0 Q_{gS} \tag{PIV.13}$$

Consequently,

$$\begin{aligned} Q_{gS} &= p_B Q_{gB}/p_0 = 248 \times 10^3 \times 3.0 \times 10^{-4}/(101.3 \times 10^3) \\ &= 7.34 \times 10^{-4}\,\mathrm{m}^3/\mathrm{s} \end{aligned} \tag{PIV.14}$$

Such a big pool can be observed in Kaiyukan (one of the biggest aquaria in Japan) in Osaka Bay Area. A ring vortex was observed, generated by injecting air into the pool through an annular bottom nozzle expanded in the course of rising due to a decrease in the static pressure around it. Finally, it broke up into many smaller bubbles due to hydrodynamic instability.

Problem IV.5 There are five particles of 1, 1.5, 2.0, 2.5, and 3.0 cm in diameter. Determine the mean diameters: D_{10} through D_{32}.

Problem IV.6 Air is injected into a water bath through a single-hole bottom nozzle at a flow rate, Q_g, of 3.0×10^{-4} m^3/s. The bath diameter, D, is 2.00×10^{-1} m. The inner and outer diameters of the nozzle, d_{ni} and d_{no}, are 2.0×10^{-3} m and 4.0×10^{-3} m, respectively. Calculate the frequency of bubble formation, f_{BF}, and bubble diameter, d_B, in the following two cases on the basis of empirical equation proposed by Davidson and Amick:

(a) The nozzle is wetted by water.
(b) The nozzle is poorly wetted by water.

(Ans: (a) $f_{BF} = 52.1$ Hz, $d_B = 2.22 \times 10^{-2}$ m, (b) $f_{BF} = 38.6$ Hz, $d_B = 2.46 \times 10^{-2}$ m)

Problem IV.7 Air is injected into a water bath through a single-hole bottom nozzle at a flow rate, Q_g, of 3.0×10^{-4} m^3/s. The bath diameter, D, is 2.00×10^{-1} m. The inner and outer diameters of the nozzle, d_{ni} and d_{no}, are 2.0×10^{-3} m and 4.0×10^{-3} m, respectively. Calculate the frequency of bubble formation, f_{BF}, and bubble diameter, d_B, in the following two cases on the basis of empirical equation proposed by Iguchi et al. The densities of air and water, ρ_g and ρ_L, are 1.22 kg/m^3 and 997 kg/m^3, respectively. The surface tension of water, σ, is 73 mN/m.

(a) The nozzle is wetted by water.
(b) The nozzle is poorly wetted by water.

(Ans: (c) f_{BF} =58.4 Hz, $d_B = 2.14 \times 10^{-2}$ m, (d) $f_{BF} = 41.3$ Hz, $d_B = 2.40 \times 10^{-2}$ m)

Index

M. Iguchi and O.J. Ilegbusi, *Basic Transport Phenomena in Materials Engineering*,
DOI 10.1007/978-4-431-54020-5, © Springer Japan 2014

MIX
Papier aus verantwortungsvollen Quellen
Paper from responsible sources
FSC® C105338

If you have any concerns about our products,
you can contact us on
ProductSafety@springernature.com

In case Publisher is established outside the EU,
the EU authorized representative is:
Springer Nature Customer Service Center GmbH
Europaplatz 3, 69115 Heidelberg, Germany

Printed by Libri Plureos GmbH
in Hamburg, Germany